Praise for Surviva

"[Douglas] Rushkoff is well worth reading [and] uncannily right."
—Michele Pridmore-Brown, *Times Literary Supplement* (UK)

"Harrowing and illuminating." —Chris Barsanti, *PopMatters*

"Intriguing. . . . [*Survival of the Richest*] shows the degree to which serious money is fretting about a looming disaster [and how] this scramble to organise the logistics of bunker life may make the underlying problems worse."
—Gillian Tett, *Financial Times* (UK)

"As tech giants amass unfathomable wealth, looking to the stars or their bunkers for escape from the fires they have lit, what are the rest of us to do? A devastating portrait of the cultures and logics underlying big tech. Rushkoff is going to make you mad enough to fight back. A vital, lucid, and enraging read."
—Naomi Klein, author of
This Changes Everything: Capitalism vs. the Climate

"*Survival of the Richest* is more than a primer on a soulless world-view pervading all aspects of life. . . . Rushkoff offers something at once more realistic and more imaginative: mutual regard, responsibility, and flourishing. In so doing, he mounts an impassioned defense of everything and everyone marked expendable in the fanatical pursuit of a blank slate." —Jenny Odell, author of
How To Do Nothing: Resisting the Attention Economy

"There are plenty of books decrying the horrors of twenty-first-century monopoly capitalism, but none quite like *Survival of the*

Richest. Rushkoff is essential—not just a passionate visionary on the side of the angels, but the rare one who can write."

—Kurt Andersen, author of
Evil Geniuses: The Unmaking of America: A Recent History

"[Rushkoff's] report is both fierce and amazed in the face of capitalism's delusions; I for one am sharpening my pitchfork."
—Jonathan Lethem, author of *The Ecstasy of Influence: Nonfictions, Etc.*

"A sober, scathing oddsmaking on the recursive wager of the ultrarich: that they can insulate themselves from the world they're creating in their rush to insulate themselves from the world they're creating." —Cory Doctorow, cofounder of BoingBoing

"A hilarious and lacerating look at the elite sociopathy wrecking the world, and a call to arms for how the rest of us can fight it."
—Molly Crabapple, author of *Drawing Blood*

"Beyond eye-opening, this book is eye-popping. A master storyteller, Rushkoff brings to life perhaps the greatest challenge of our time. A must-read."
—Frances Moore Lappé, author of *Diet for a Small Planet*

"Douglas Rushkoff's keen eye as a seasoned media analyst, combined with his flair and wit as a writer and a performer, shine in this book. Rushkoff confronts the reader with a ridiculous conundrum: how is it possible that people who have powerfully shaped our society and economy and have reaped enormous financial rewards in the process are doing everything possible to escape the world they've created?"
—Marina Gorbis, executive director of Institute for the Future

"With razor-sharp insight, Rushkoff unwraps the dazzling facade of the technological dream, revealing the alarming Mindset that underlies promises of planetary salvation. . . . Ultimately, Rushkoff demonstrates, the growth-based techno-solutionism inspired by the Mindset will drive our civilization toward collapse unless we begin to recognize capitalism as the underlying issue that needs to be addressed."

—Jeremy Lent, author of *The Patterning Instinct: A Cultural History of Humanity's Search for Meaning*

Survival of the Richest

ESCAPE FANTASIES OF
THE TECH BILLIONAIRES

DOUGLAS RUSHKOFF

W. W. NORTON & COMPANY
Celebrating a Century of Independent Publishing

For Mark Filippi, Michael Nesmith, and
Genesis Breyer P-Orridge. Wish you were here.

For information about permission to reproduce selections from this book, write to Permissions, W. W. Norton & Company, Inc., 500 Fifth Avenue, New York, NY 10110

For information about special discounts for bulk purchases, please contact W. W. Norton Special Sales at specialsales@wwnorton.com or 800-233-4830

Manufacturing by Lakeside Book Company
Book design by Chris Welch
Production manager: Anna Oler

Library of Congress Cataloging-in-Publication Data

Names: Rushkoff, Douglas, author.
Title: Survival of the richest : escape fantasies of the tech billionaires / Douglas Rushkoff.
Description: First edition. | New York, NY : W. W. Norton & Company, Inc., [2022] |
    Includes bibliographical references.
Identifiers: LCCN 2022027182 | ISBN 9780393881066 (cloth) | ISBN 9780393881073 (epub)
Subjects: LCSH: Technology and civilization—Moral and ethical aspects. | Survivalism—
    Moral and ethical aspects. | Billionaires—Conduct of life.
Classification: LCC T14.5 .R863 2022 | DDC 303.48/3—dc23/eng/20220819
LC record available at https://lccn.loc.gov/2022027182

ISBN 978-1-324-06606-4 pbk.

W. W. Norton & Company, Inc., 500 Fifth Avenue, New York, N.Y. 10110
www.wwnorton.com

W. W. Norton & Company Ltd., 15 Carlisle Street, London W1D 3BS

1 2 3 4 5 6 7 8 9 0

Contents

Survival of the Richest

Meet The Mindset

I got invited to a super-deluxe resort to deliver a speech to what I assumed would be a hundred or so investment bankers. It was by far the largest fee I had ever been offered for a talk—about a third of my annual salary as a professor at a public college—all to deliver some insight on "the future of technology."

As a humanist who writes about the impact of digital technology on our lives, I am often mistaken for a futurist. And I've never really liked talking about the future, especially for wealthy people. The Q & A sessions always end up more like parlor games, where I'm asked to opine on the latest technology buzzwords as if they were ticker symbols on a stock exchange: AI, VR, CRISPR. The audiences are rarely interested in learning about how these technologies work or their impact on society beyond the binary choice of whether or not to invest in them. But money talks, and so do I, so I took the gig.

I flew business class. They gave me noise-canceling headphones to wear and warmed mixed nuts to eat (yes, they *heat* the nuts) as I composed a lecture on my MacBook about how digital businesses could foster circular economic principles rather than doubling down on extractive growth-based capitalism—painfully aware that neither the ethical value of my words nor the carbon offsets I had purchased along with my ticket could possibly compensate for the environmental damage I was doing. I was funding my mortgage and my daughter's college savings plan at the expense of the people and places down below.

A limo was waiting for me at the airport and took me straight out into the high desert. I tried to make conversation with the driver about the UFO cults that operate in that part of the country and the desolate beauty of the terrain compared with the frenzy of New York. I suppose I felt an urge to make sure he understood I'm not of the class of people who usually sit in the back of a limo like this. As if to make the opposite point about himself, he finally disclosed that he wasn't a full-time driver but a day trader a bit down on his luck after a few "poorly timed puts."

As the sun began to dip over the horizon, I realized I had been in the car for three hours. What sort of wealthy hedge fund types would drive this far from the airport for a conference? Then I saw it. On a parallel path next to the highway, as if racing against us, a small jet was coming in for a landing on a private airfield. Of course.

Just over the next bluff was the most luxurious yet isolated place I've ever been. A resort and spa in the middle of, well, nowhere. A scattering of modern stone and glass structures were nestled into a big rock formation, looking out on the infinity of the desert. I saw no one but attendants as I checked in and had to use a map to find my way to my private "pavilion" for the night. I had my own outdoor hot tub.

The next morning, two men in matching Patagonia fleece came for me in a golf cart and conveyed me through rocks and underbrush to a meeting hall. They left me to drink coffee and prepare in what I figured was serving as my green room. But instead of me being wired with a microphone or taken to a stage, my audience was brought in to me. They sat around the table and introduced themselves: five super-wealthy guys—yes, all men—from the upper echelon of the tech investing and hedge fund world. At least two of them were billionaires. After a bit of small talk, I realized they had no interest in the talk I had prepared about the future of technology. They had come to ask questions.

They started out innocuously and predictably enough. Bitcoin or Ethereum? Virtual reality or augmented reality? Who will get quantum computing first, China or Google? But they didn't seem to be taking it in. No sooner would I begin to explain the merits of proof-of-stake versus proof-of-work blockchains than they would move to the next question. I started to feel like they were testing me—not my knowledge so much as my scruples.

Eventually, they edged into their real topic of concern: New Zealand or Alaska? Which region will be less impacted by the coming climate crisis? It only got worse from there. Which was the greater threat: climate change or biological warfare? How long should one plan to be able to survive with no outside help? Should a shelter have its own air supply? What is the likelihood of groundwater contamination? Finally, the CEO of a brokerage house explained that he had nearly completed building his own underground bunker system, and asked, "How do I maintain authority over my security force after the event?" The Event. That was their euphemism for the environmental collapse, social unrest, nuclear explosion, solar storm, unstoppable virus, or malicious computer hack that takes everything down.

This single question occupied us for the rest of the hour. They knew armed guards would be required to protect their compounds from raiders as well as angry mobs. One had already secured a dozen Navy SEALs to make their way to his compound if he gave them the right cue. But how would he pay the guards once even his crypto was worthless? What would stop the guards from eventually choosing their own leader?

The billionaires considered using special combination locks on the food supply that only they knew. Or making guards wear disciplinary collars of some kind in return for their survival. Or maybe building robots to serve as guards and workers—if that technology could be developed "in time."

I tried to reason with them. I made pro-social arguments for partnership and solidarity as the best approaches to our collective, long-term challenges. The way to get your guards to exhibit loyalty in the future is to treat them like friends right now, I explained. Don't just invest in ammo and electric fences, invest in people and relationships. They rolled their eyes at what must have sounded to them like hippie philosophy, so I cheekily suggested that the way to make sure your head of security doesn't slit your throat tomorrow is to pay for his daughter's bat mitzvah today. They laughed. At least they were getting their money's worth in entertainment.

I could tell they were also a bit annoyed. I wasn't taking them seriously enough. But how could I? This was probably the wealthiest, most powerful group I had ever encountered. Yet here they were, asking a Marxist media theorist for advice on where and how to configure their doomsday bunkers. That's when it hit me:

at least as far as these gentlemen were concerned, this *was* a talk about the future of technology.

Taking their cue from Tesla founder Elon Musk colonizing Mars, Palantir's Peter Thiel reversing the aging process, or artificial intelligence developers Sam Altman and Ray Kurzweil uploading their minds into supercomputers, they were preparing for a digital future that had less to do with making the world a better place than it did with transcending the human condition altogether. Their extreme wealth and privilege served only to make them obsessed with insulating themselves from the very real and present danger of climate change, rising sea levels, mass migrations, global pandemics, nativist panic, and resource depletion. For them, the future of technology is about only one thing: escape from the rest of us.

These people once showered the world with madly optimistic business plans for how technology might benefit human society. Now they've reduced technological progress to a video game that one of them wins by finding the escape hatch. Will it be Bezos migrating to space, Thiel to his New Zealand compound, or Zuckerberg to his virtual Metaverse? And these catastrophizing billionaires are the presumptive *winners* of the digital economy—the supposed champions of the survival-of-the-fittest business landscape that's fueling most of this speculation to begin with.

Of course, it wasn't always this way. There was a brief moment, in the early 1990s, when the digital future felt open-ended. In spite of its origins in military cryptography and defense networking, digital technology had become a playground for the counterculture, who saw in it the opportunity to invent a more inclusive, distributed, and participatory future. Indeed, the "digital renaissance," as I began to call it back in 1991, was about the unbridled potential of the collective human imagination. It spanned everything from chaos math and quantum physics to fantasy role-playing.

Many of us in that early cyberpunk era believed that—connected and coordinated as never before—human beings could create any future we imagined. We read magazines called *Reality Hackers*, *FringeWare*, and *Mondo2000*, which equated cyberspace with psychedelics, computer hacking with conscious evolution, and online networking with massive electronic dance music parties called raves. The artificial boundaries of linear, cause-and-effect reality and top-down classifications would be superseded by a fractal of emerging interdependencies. Chaos was not random, but rhythmic. We would stop seeing the ocean through the cartographer's grid of latitude and longitude lines, but in the underlying patterns of the water's waves. "Surf's up," I announced in my first book on digital culture.

No one took us very seriously. That book was actually canceled by its original publisher in 1992 because they thought the computer networking fad would be "over" before my publication date in late 1993. It wasn't until *Wired* magazine launched later that year, reframing the emergence of the internet as a business opportunity, that people with power and money began to take notice. The fluorescent pages of the magazine's first issue announced that "a tsunami was coming." The articles suggested that only the investors who kept track of the scenario-planners and futurists on their pages would be able to survive the wave.

This wasn't going to be about the psychedelic counterculture, hypertext adventures, or collective consciousness. No, the digital revolution wasn't a revolution at all but a business opportunity—a chance to pump steroids into the already dying NASDAQ stock exchange, and maybe to milk another couple of decades of growth out of an economy presumed dead since the biotech crash of 1987.

Everyone crowded back into the tech sector for the dotcom boom. Internet journalism moved off the culture and media pages of the newspaper and into the business section. Established business interests saw new potentials in the net, but only for the same old extraction they'd always done, while promising young technologists were seduced by unicorn IPOs and multimillion-dollar payouts. Digital futures became understood more like stock futures or cotton futures—something to predict and make bets on. Likewise, technology users were treated less as creators to empower than consumers to manipulate. The more predictable the users' behaviors, the more certain the bet.

Nearly every speech, article, study, documentary, or white paper on the emerging digital society began to point to a ticker symbol. The future became less a thing we create through our present-day choices or hopes for humankind than a predestined scenario we bet on with our venture capital but arrive at passively.

This freed everyone from the moral implications of their activities. Technology development became less a story of collective flourishing than personal survival through the accumulation of wealth. Worse, as I learned in writing books and articles about such compromises, to call attention to any of this was to unintentionally cast oneself as an enemy of the market or an anti-technology curmudgeon. After all, the growth of technology and that of the market were understood as the same thing: inevitable, and even morally desirable.

Market sensibilities overpowered much of the media and intellectual space that would have normally been filled by a consideration of the practical ethics of impoverishing the many in the name of the few. Too much mainstream debate centered instead on abstract hypotheticals about our predestined high-tech future: Is it fair for a stock trader to use smart drugs? Should children get implants for foreign languages? Do we want autonomous vehicles to prioritize the lives of pedestrians over those of its passengers? Should the first Mars colonies be run as democracies? Does changing my DNA undermine my identity? Should robots have rights?

Asking these sorts of questions, which we still do today, may be philosophically entertaining. But it is a poor substitute for wrestling with the real moral quandaries associated with unbridled technological development in the name of corporate capitalism. Digital platforms have turned an already exploitative and extractive marketplace (think Walmart) into an even more dehumanizing successor (think Amazon). Most of us became aware of these downsides in the form of automated jobs, the gig economy, and the demise of local retail along with local journalism.

But the more devastating impacts of pedal-to-the-metal digital capitalism fall on the environment, the global poor, and the civilizational future their oppression portends. The manufacture of our computers and smartphones still depends on networks of slave labor. These practices are deeply entrenched. A company called Fairphone, founded to make and market ethical phones, learned it was impossible. (The company's founder now sadly refers to its products as "fairer" phones.) Meanwhile, the mining of rare earth metals and disposal of our highly digital technologies destroys human habitats, replacing them with toxic waste dumps, which are then picked over by impoverished indigenous children and their families, who sell usable materials back to the manufacturers—who then cynically claim this "recycling" is part of their larger efforts at environmentalism and social good.

This "out of sight, out of mind" externalization of poverty and poison doesn't go away just because we've covered our eyes with VR goggles and immersed ourselves in an alternate reality. If anything, the longer we ignore the social, economic, and environmental repercussions, the more of a problem they become. This, in turn, motivates even more withdrawal, more isolationism and apocalyptic fantasy—and more desperately concocted technologies and business plans. The cycle feeds itself.

The more committed we are to this view of the world, the more we come to see other human beings as the problem and technology as the way to control and contain them. We treat the deliciously quirky, unpredictable, and irrational nature of humans less as a feature than a bug. No matter their own embedded biases, technologies are declared neutral. Any bad behaviors they induce in us are just a reflection of our own corrupted core. It's as if some innate, unshakeable human savagery is to blame for our troubles. Just as the inefficiency of a local taxi market can be "solved" with an app that bankrupts human drivers, the vexing inconsistencies of the human psyche can be corrected with a digital or genetic upgrade.

Ultimately, according to the technosolutionist orthodoxy, the human future climaxes by uploading our consciousness to a computer or, perhaps better, accepting that technology itself is our evolutionary successor. Like members of a gnostic cult, we long to enter the next transcendent phase of our development, shedding our bodies and leaving them behind, along with our sins and troubles, and—most of all—our economic inferiors.

Our movies and television fare play out these fantasies for us. Zombie shows depict a post-apocalypse where people are no better than the undead—and seem to know it. Worse, these shows invite viewers to imagine the future as a zero-sum battle between the remaining humans, where one group's survival is dependent on another one's demise. Even our most forward-thinking science fiction shows now depict robots as our intellectual and ethical superiors. It's always the humans who are reduced to a few lines of code, and the artificial intelligences who learn to make more complex and willful choices.

The mental gymnastics required for such a profound role reversal between humans and machines all depend on the underlying assumption that most humans are essentially worthless and unthinkingly self-destructive. Let's either change them or get away from them, forever. Thus, we get tech billionaires launching electric cars into space—as if this symbolizes something more than one billionaire's capacity for corporate promotion. And if a few people do reach escape velocity and somehow survive in a bubble on Mars—despite our inability to maintain such a bubble even here on Earth in either of two multibillion-dollar Biosphere trials—the result would be less a continuation of the human diaspora than a lifeboat for the elite. Most thinking, breathing human beings understand there is no escape.

What I came to realize as I sat sipping imported iceberg water and pondering doomsday scenarios with our society's great winners is that these men are actually the losers. The billionaires who called me out to the desert to evaluate their bunker strategies are not the victors of the economic game so much as the victims of its perversely limited rules. More than anything, they have succumbed to a mindset where "winning" means earning enough money to insulate themselves from the damage they are creating by earning money in that way. It's as if they want to build a car that goes fast enough to escape from its own exhaust.

Yet this Silicon Valley escapism—let's call it The Mindset—encourages its adherents to believe that the winners can somehow leave the rest of us behind. Maybe that's been their objective all along. Perhaps this fatalist drive to rise above and separate from humanity is no more the result of runaway digital capitalism than its cause—a way of treating one another and the world that can be

traced back to the sociopathic tendencies of empirical science, individualism, sexual domination, and perhaps even "progress" itself.

Yet while tyrants since the time of Pharaoh and Alexander the
Great may have sought to sit atop great civilizations and rule them
from above, never before have our society's most powerful players
assumed that the primary impact of their own conquests would
be to render the world itself unlivable for everyone else. Nor have
they ever before had the technologies through which to program
their sensibilities into the very fabric of our society. The landscape
is alive with algorithms and intelligences actively encouraging
these selfish and isolationist outlooks. Those sociopathic enough
to embrace them are rewarded with cash and control over the rest
of us. It's a self-reinforcing feedback loop. This is new.

Amplified by digital technologies and the unprecedented wealth
disparity they afford, The Mindset allows for the easy externalization of harm to others, and inspires a corresponding longing for
transcendence and separation from the people and places that have
been abused. As we will see, The Mindset is based in a staunchly
atheistic and materialist scientism, a faith in technology to solve
problems, an adherence to biases of digital code, an understanding
of human relationships as market phenomena, a fear of nature and
women, a need to see one's contributions as utterly unique innovations without precedent, and an urge to neutralize the unknown
by dominating and de-animating it.

Instead of just lording over us forever, however, the billionaires
at the top of these virtual pyramids actively seek the endgame. In
fact, like the plot of a Marvel blockbuster, the very structure of The
Mindset *requires* an endgame. Everything must resolve to a one or
a zero, a winner or loser, the saved or the damned. Actual, immi-
nent catastrophes from the climate emergency to mass migrations
support the mythology, offering these would-be superheroes the

opportunity to play out the finale in their own lifetimes. For The Mindset also includes a faith-based Silicon Valley certainty that they can develop a technology that will somehow break the laws of physics, economics, and morality to offer them something even better than a way of saving the world: a means of escape from the apocalypse of their own making.

The Insulation Equation

BILLIONAIRE BUNKER STRATEGIES

By the time I boarded my return flight to New York, my mind was reeling with the implication of The Mindset. Where had it come from? What caused it? What were its main tenets? Who were its true believers? What, if anything, could we do to resist it? Before I had even landed, I posted an article about my strange encounter—to surprising effect.

Almost immediately, I began receiving inquiries from businesses catering to the billionaire prepper, all hoping I would make some introductions on their behalf to the five men I had written about. I heard from a real estate agent who specializes in disaster-proof listings, a company taking reservations for its third underground dwellings project, and a security firm offering various forms of "risk management."

But the message that got my attention came from a former president of the American Chamber of Commerce in Latvia. J. C. Cole had witnessed the fall of the Soviet empire as well as what it took to rebuild a working society almost from scratch. He had also served as landlord for the American and European Union embassies, and learned a whole lot about security systems and evacuation plans. "You certainly stirred up a bee's nest," he began his first email to me. "I find it quite accurate—the wealthy hiding in their bunkers will have a problem with their security teams . . . I believe you are correct with your advice to 'treat those people really well, right now,' but also the concept may be expanded and I believe there is a better system that would give much better results."

He proceeded to lay out the facts. He felt certain that the "Event"—a gray swan, or predictable catastrophe triggered by our enemies, Mother Nature, or just by accident—was inevitable. He had done a SWOT analysis—Strengths, Weaknesses, Opportunities, and Threats—on the situation, and concluded that preparing for calamity requires us to take the very same measures as trying to prevent one. "By coincidence," he explained, "I am setting up a series of Safe Haven Farms in the NYC area. These are designed to best handle an 'event' and also benefit society as semi-organic farms. Both within three hours' drive from the City—close enough to get there when it happens."

I couldn't resist. Here was a prepper with security clearance, field experience, and food sustainability expertise. He believed the best way to cope with the impending disaster was to change the way we treat one another, the economy, and the planet right now—while also developing a network of secret, totally self-sufficient residential farm communities for millionaires, guarded by Navy SEALs armed to the teeth.

J.C. is currently developing two farms as part of his Safe Haven project. Farm 1, outside Princeton, is his show model and "works well as long as the Thin Blue Line is working." The second one, somewhere in the Poconos, has to remain a secret. "The fewer people who know the locations, the better," he explained, along with a link to the *Twilight Zone* episode where panicked neighbors break into a family's bomb shelter during a nuclear scare. "The primary value of Safe Haven is Operational Security, nicknamed OpSec by the military. If/when the supply chain breaks, the people will have no food delivered. Covid-19 gave us the wake-up call as people started fighting over TP. When it comes to a shortage of food it will be vicious. That is why those intelligent enough to invest have to be stealth."

J.C. offered to come into New York to show me his proposal, but I wanted to see the real thing. He was delighted, and invited me down to New Jersey. "Wear boots," he said. "The ground is still wet." Then he asked, "Do you shoot?"

The farm itself was serving as an equestrian center and tactical training facility in addition to raising goats and chickens. J.C. showed me how to hold and shoot a Glock at a series of outdoor targets shaped like bad guys, while he grumbled about the way Senator Diane Feinstein had arbitrarily limited the number of rounds one could legally fit in a magazine for the handgun. J.C. knew his stuff. I asked him about various combat scenarios. How does one defend against a whole gang of thugs invading one's farm? "You don't," he said. "The bottom line of prepping is to get away."

Of course, if you have a compound like the one J.C. was building, things are a little different. "The only way to protect your family is with a group," he said. That's really the whole point of his project—to gather a team capable of sheltering in place for a year or more, while also defending itself from those who haven't prepared. "The SWAT team of a city police force have visited here. They all said they'd be here at the first sign of trouble." J.C. is also hoping to train young farmers in sustainable agriculture, and to secure at least one doctor and dentist for each location.

We had to finish shooting before a teenager arrived to practice jumping with her horse. On the way back to the main building, J.C. showed me the "layered security" protocols he had learned designing embassy properties: a fence around the whole place, NO TRESPASSING signs, guard dogs, surveillance cameras . . . all disincentives meant to prevent a violent confrontation. He paused for a minute as he stared down the drive. "Honestly, I am less concerned about gangs with guns than the woman at the end of the driveway holding a baby and asking for food." He paused, and sighed, "I don't want to be in that moral dilemma."

That's why J.C.'s real passion isn't just to build a few isolated, militarized retreat facilities for millionaires, but to prototype locally owned sustainable farms that can be modeled by others and ultimately help restore regional food security in America. The "just-in-time" delivery system preferred by agricultural conglomerates renders most of the nation vulnerable to a crisis as minor as a power outage or transportation shutdown. Meanwhile, the centralization of the agricultural industry has left most farms utterly dependent on the same long supply chains as urban consumers. "Most egg farmers can't even raise chickens," J.C. explained as he showed me his henhouses. "They buy chicks. I've got roosters."

J.C. is no hippie environmentalist. He refers to Hillary Clinton only as "her" and publishes pieces online about America's deep state misadventures and the coming oil wars. But his business model is based in the same communitarian spirit I tried to convey to the billionaires: the way to keep the hungry hordes from storming the gates is by getting them food security now. So for three million dollars, investors not only get a maximum security compound in which to ride out the coming plague, solar storm, or electric grid collapse. They also get a stake in a potentially profitable network of local farm franchises that could reduce the probability of a catastrophic event in the first place. His business would do its best to ensure there are as few hungry children at the gate as possible when the time comes to lock down.

So far, J. C. Cole has been unable to convince anyone to invest in American Heritage Farms. That doesn't mean no one is investing in such schemes. It's just that the ones that attract more attention and cash don't generally have these cooperative components. They're more for people who want to go it alone. Most billionaire preppers don't want to have to learn to get along with a community of farmers or, worse, spend their winnings funding a national food resilience program. The mindset that requires safe havens is less concerned with preventing moral dilemmas than simply keeping them out of sight.

Many of those seriously seeking a safe haven simply hire one of several prepper construction companies to bury a prefab steel-lined bunker somewhere on one of their existing properties. Rising S Company out of Texas builds and installs bunkers and tornado shelters for as little as $40,000 for an eight-by-twelve-foot emergency hideout all the way up to the $8.3 million Luxury Series "Aristocrat," complete with pool and bowling lane. While they've got photos of the lower-priced models on their website, the larger ones are depicted in virtual walkthroughs, likely because not many (if any) are actually being constructed on that scale. These are pretty spartan facilities, anyway—more like repurposed shipping containers than James Bond-level fantasy hideouts. The enterprise originally catered to families seeking temporary storm shelters, before it went into the long-term apocalypse business. The company logo, complete with three crucifixes, suggests their services are geared more toward Christian evangelist preppers in red state America than billionaire tech bros playing out sci-fi scenarios.

There's something much more whimsical about the facilities in which most of the billionaires—or, more accurately, aspiring billionaires—actually invest. A company called Vivos is selling luxury underground apartments in converted Cold War munitions storage facilities, missile silos, and other fortified locations around the world. Like miniature Club Med resorts, they offer private suites for individuals or families, and larger common areas with pools, games, movies, and dining. Ultra-elite shelters like the Oppidum in the Czech Republic claim to cater to the billionaire class, and pay more attention to the long-term psychological health of residents. They provide imitation of natural light, such as a pool with a simulated sunlit garden area, a wine vault, and other amenities to make the wealthy feel at home.

On closer analysis, however, the probability of a fortified bunker actually protecting its occupants from the reality of, well, reality, is very slim. For one, the closed ecosystems of underground facilities are preposterously brittle. The diversity found in genuine, real-world biomes cushions them and their inhabitants from catastrophe. In nature, a disease, drought, or invader may threaten one species yet be successfully mitigated by another. An indoor, sealed hydroponic garden is vulnerable to contamination. Vertical farms with moisture sensors and computer-controlled irrigation systems look great in business plans and on the rooftops of Bay Area start-ups; when a palette of topsoil or a row of crops goes wrong, it can simply be pulled and replaced. The hermetically sealed apocalypse "grow room" doesn't allow for such do-overs.

Just the *known* unknowns are enough to dash any reasonable hope of survival. But this doesn't seem to stop wealthy preppers from trying. The *New York Times* reported that real estate agents specializing in private islands were overwhelmed with inquiries during the Covid-19 pandemic. Prospective clients were even asking about whether there was enough land to do some agriculture in addition to installing a helicopter landing pad. But while a private island may be a good place to wait out a temporary plague, turning it into a self-sufficient, defensible ocean fortress is harder than it sounds. Small islands are utterly dependent on air and sea deliveries for basic staples. Solar panels and water filtration equipment need to be replaced and serviced at regular intervals. The billionaires who reside in such locales are more, not less, dependent on complex supply chains than those of us embedded in industrial civilization.

Not that the environment is truly sealed, anyway. Everything gets everywhere. Toxic clouds, plague, and radiation have a way of spreading and seeping through the most well-thought-out barricades. HEPA filters need to be regularly replaced, and sometimes fail even when they are. Air pollution from factories in China and forest fires in Europe and California already travels to distant continents, measurably contaminating Everest and Katmandu. Cancer-causing microplastics are as plentiful in the polar ice as they are in the typical European town. The average American already consumes about a credit card worth of plastic a month, according to a World Wide Fund for Nature study. Just read the news. There is no escape.

Surely the billionaires who brought me out for advice on their exit strategies were aware of these limitations. Could it have all been some sort of game? Five men sitting around a poker table, each wagering his escape plan was best? Was I supposed to be playing the part of the neutral dealer, or the fantasy role-playing game master, meting out judgment on each of the scenarios they described?

Still, there was something more going on here, as well. If they were in it just for fun, they wouldn't have called for me. They would have flown out the author of a zombie apocalypse comic book. If they wanted to water-test their bunker plans, they'd have hired a security expert from Blackwater or the Pentagon. They seemed to want something more. Their language went far beyond questions of disaster preparedness and verged into politics and philosophy: words like individuality, sovereignty, governance, and autonomy.

That's because it wasn't their actual bunker strategies I had been brought out to evaluate so much as the philosophy and mathematics they were using to justify their commitment to escape. They were working out what I've come to call the Insulation Equation: could they earn enough money to insulate themselves from the reality they were creating by earning money in this way? Was there any valid justification for striving to be so successful that they could simply leave the rest of us behind—apocalypse or not?

As holders of The Mindset, they have been rejecting the collective polity all along, and embracing the hubristic notion that with enough money and technology, the world can be redesigned to one's personal specifications. Their various self-sovereignty escape initiatives amount to the same techno-libertarian world-building fantasy exemplified by the ultra-billionaire's competition to colonize Mars, but designed for implementation right here on Earth. Only trillionaires will actually make it into space to terraform planets, anyway. The cohort who solicited my doomsday advice readily admitted they were "low-level billionaires" who could at best hitch a ride with Elon Musk, Richard Branson, or Jeff Bezos— who are themselves still at least a few generations away from colonizing anything.

Offering a slightly more reasonable techno-utopian escape fantasy, the "seasteading" movement—publicized in a flurry of maga-

zine stories a few years ago—promises a sustainable solution to a
world of climate catastrophe, social chaos, and economic collapse.
In the Minecraft-meets-Waterworld future envisioned by "aquapreneurs," wealthy people are to live in independent, free-floating
city-states—giant clusters of high-tech rafts using clean, renewable ocean thermal energy to power themselves and escape from
a civilization of oil-drilling land dwellers. The hype around these
initiatives may have died down, but several billionaires and even
some legitimate organizations including the United Nations and
MIT are still hard at work on humanity's return to the sea.

Proponents of seasteading seem to start every conversation
with the promise of sustainability, environmentalism, or insulation from risks like Covid or climate chaos (why fear rising oceans
if you're already living on the ocean?) Eventually, however, they
always get to more ideological motivations for leaving terra firma
behind. The mission statement of the Seasteading Institute puts it
plainly: "To establish permanent, autonomous ocean communities to enable experimentation and innovation with diverse social,
political, and legal systems."

The tech entrepreneurs investing in these ocean schemes seek
to retrieve the Wild West free-for-all associated with the early
internet. It has little to do with the water and everything to do
with political autonomy—freedom to live ruled only by The Mindset. Unfettered and unregulated by the backwards thinking of
nation-states, aquapreneurs will be free to reimagine civilization
as an ultra-libertarian experiment. They will rapidly prototype
new forms of government, and determine what—if any—nods
to civics or collectivism are even necessary. As the Seasteading
Institute website explains, "We've had the agricultural revolution,
the commercial and industrial revolutions, but why not a governance revolution? Enter the sea." The ocean will be the means to

an end—a way of redefining one's sovereignty from the bottom up, being always absolutely in charge of one's own personal allegiance, expression of values, and obligation to the law.

It is a vision for something like a global un-conference, where each individual or family builds or buys their own high-tech floating villa or "nano-nation," and then floats to whichever modular cluster-nation offers the best system of government. If you stop liking the way the government is operating, you simply disconnect and propel yourself to another cluster, somewhere else in the ocean. In a free market free-for-all, startup societies will compete for inhabitants much like social networks compete for users or Burning Man camps compete for visitors. Moreover, free of national regulations, aquapreneurs will be able to develop technologies and make scientific breakthroughs impossible in countries imposing legal or moral restrictions on genetic engineering, cloning, or nanotechnology.

Shrouded in the urgency of environmentalism and the optimism of technology innovation, self-sovereignty fantasies like this betray the underlying urge among the techno-libertarian elite to stop submitting to congressional inquiries, anti-monopoly regulations, or regressive technophobia, and to take their ball and go play elsewhere.

Whether on land, on sea, or in outer space, the quest for self-sovereignty is less important as an example of apocalypse preparedness than it is as an exposé of the underlying, Ayn Rand fantasies of the tech elite: the most rational and productive among us escape to pursue their self-interests, empowered to build an independent economy of their own, free from the moral consequences of their actions.

Mergers and Acquisitions

ALWAYS HAVE AN EXIT STRATEGY

This is not how most of us thought digital technology would change human culture.

My own first exposure to computers was in ninth grade. Our district's Board of Education had purchased its first IBM mainframe to handle school records and, as an afterthought, they installed three terminals in the math department office for interested students. Of course, the kids quickly learned more about the school's computer system than the adult administrators. On a daily basis, computer nerds would get called out of class to go fix the system that coordinated teachers' paychecks and tabulated students' grades. Their whole microculture had an ethos of serving others, teaching newbies, and sharing everything.

By necessity, that's the ethos that ruled in the greater computer culture as well. People didn't own personal computers, yet. They worked at terminals connected to big computers somewhere else. This meant that everyone had to share computing resources—the limited cycles of the machines. Much of our early software simply orchestrated all that sharing, such as what partition of a hard drive would house each user's "directory" of files, or what time someone might be allowed to "run" their program on the central processor.

Naturally, the software written by these users was "shareware" as well. Why would someone even think to charge for a program they'd written? The measure of a program's success was how widely it was used. It was a point of pride. Money didn't enter into it. In fact, back then, parents would actually *worry* when their children got interested in computers. It looked as if we were throwing our lives away to play video games and never earn a real living. They had no idea what was really going on.

I only came to appreciate the immensity of what was occurring, myself, when I got to college and visited the computer lab to type up my senior thesis. I'll always remember the moment when the grad student working there showed me how to save my work. She told me I could save my file as "read only," meaning that others could read my file, or "read write," meaning that they could "write to" or *change* the file I had saved. Every file was read-only or read-write. It sounded simple at the time, but I remember leaving the lab and seeing the world differently after that. Which things in the world were read-only, and which were read-write? Why was money read-only? And religion? What if we could change them? How much of the world was arbitrarily protected from our intervention, and who got to make those decisions? I had been raised in the read-only media environment of television. A spectator. But I was coming to realize this was all about to change.

My friends moving out to Silicon Valley weren't going there just to program computers, but to program reality. They weren't the pocket protector geeks I knew from the high school math office. They were the artists and psychedelics users I knew from my music and theater classes. Deadheads, Wiccans, and Brian Eno enthusiasts. As experienced hallucinators, members of the psychedelic community were particularly well-suited to imagining virtual environments and new modes of human communication. Tech companies actively sought out programmers from this imaginative subculture, and made special accommodations for them— such as quietly giving certain employees early warning of the drug tests they were required to conduct as defense contractors.

Psychedelic heroes of the 1960s, including LSD guru Timothy Leary, former Merry Prankster Stewart Brand, and Grateful Dead lyricist John Barlow, reassured the California counterculture that the computer revolution would be characterized less by the postwar military bureaucracy or even high-tech corporations than by the "new communalists" of Haight-Ashbury, the Whole Earth Catalog, and the hot tubs of Esalen.

By the early 1990s, psychedelic and computer culture had grown indistinguishable. Software developers who wrote code for Apple during the day came home to scrape peyote buds off cactuses and trip all night. My friends at SUN Microsystems used their high-powered computers to generate fractal imagery that was projected at Dead shows. For a while, Intel even supplied experimental virtual reality gear for kids to demo at San Francisco rave parties.

If our internet had an enemy, it wasn't the corporations giving us toys to play with and paying for our time; it was the government, which used computers for war games, arrested young hackers on trumped-up charges, and sought to censor our online communication. Everyone knew that the internet was originally a network called Arpanet, designed as a decentralized system to help the military communicate even after a calamity such as a nuclear war. We wanted to get as far from that origin as possible. Meanwhile, the FBI had conducted a series of unnecessarily heavy-handed raids called Operation Sundevil, bursting into the homes of some of our favorite teen hackers—arresting them and terrifying their families, all for relatively minor and victimless infractions. Finally, the Communication Decency Act of 1996 sought to limit what we thought of as our information freedom by banning not just "obscene" materials, but anything deemed "indecent."

John Barlow's 1996 Declaration of Independence of Cyberspace expressed what many of us were feeling, and then some: "Governments of the Industrial World, you weary giants of flesh and steel, I come from Cyberspace, the new home of Mind. On behalf of the future, I ask you of the past to leave us alone. You are not welcome among us. You have no sovereignty where we gather." Government was the enemy, and had to be kept from exercising its authority over this new collective project of humankind. Most of us missed the signature line at the end, "Davos, Switzerland, February 8, 1996"—the time and place of the famous World Economic Forum, ground zero for global capitalism.

Deregulation sounded good at the time. We were just ravers and cyberpunks, paranoid about the government arresting us for drugs. We knew John Barlow as the freewheeling Grateful Dead lyricist—not in his earlier role as the libertarian who ran Dick Cheney's congressional campaign. We didn't realize that banishing government from the internet would create a free zone for corporate colonization. We hadn't yet discovered that government and business balance each other out—a bit like fungus and bacteria in the body. Get rid of one, and the other runs rampant.

We were too excited about building our new civilization to consider such imbalances. I remember participating in a workshop at the United Nations and identifying myself on my nametag as "a citizen of cyberspace." I guess I had drunk the Kool-Aid.

But the internet finally did change for me on the morning of January 10, 2000, when an editor from the *New York Times* called to ask if I could bang out an op-ed by that afternoon. I was thrilled. The culture of the net, which I had been writing about since the late 1980s, was finally mainstream enough for the Gray Lady to solicit a think piece from a fringe, cyberpunk writer like me. But the editor wanted my commentary on something slightly different: Time Warner, a venerable "old media" company, had just announced it would be acquired by America Online—an internet access company that had only been in existence for a decade. Did this mean the internet had truly arrived? Could I please explain what this merger meant, in plain English?

I didn't want to give up the opportunity, but this wasn't really my beat. I was a media and culture writer. I figured I'd wing it. The *Times* editor saw this merger as the "birth of the digital economy," and if it was, I wanted to be present and weigh in. So I analyzed the deal from my own perspective, that of a media theorist and financial naïf. A venerable conglomerate with real assets including a film library, magazine empire, theme parks, movie studios, and thousands of miles of television cable was about to merge with the guys who mailed shrink-wrapped "10 Hours Free Access" CDs to pretty much every home in America. According to the terms of the deal, the two companies would combine into one, with AOL shareholders owning 55 percent and Time Warner just 45 percent of the new company's stock.

How could America Online, a company with virtually no real assets, accomplish such a feat? Yes, AOL had accumulated a few million users, but for several months I had been hearing that the new subscriber rate was actually slowing. The only thing peaking about AOL at the turn of the century was its stock price.

Then, all of a sudden, the move made sense to the video game geek within me: AOL was spending its "in-world" fantasy money. AOL founder Steve Case was cashing in his chips, exchanging his speculative dotcom paper assets for a majority stake in a real company with real assets. Moreover, if he was choosing to do it now, it meant he believed that his AOL stock had hit its all-time high. To me, AOL's purchase of Time Warner suggested that what then-Fed chair Alan Greenspan had called the "irrational exuberance" surrounding technology stocks had climaxed: the dotcom boom was ending.

So I wrote all this down and sent it to the editor by the 3 pm deadline. An hour later, the phone rang. "We can't run this!" he said. "Everybody is saying the deal is a great thing, and you're trying to argue that it's going to *fail*? That it's some kind of bait and switch?"

"Yeah," I replied. "Time Warner is getting screwed."

"Well, we can't run a negative piece about a merger that everyone in the Business section says is unequivocally great for old and new media alike."

"Maybe you should have someone from Business write it, then."

They did. Along with hundreds of other editorials from business experts, the *New York Times*'s op-ed page extolled the virtues of the deal. I got the piece I wrote placed in the *Guardian*, where contrarian views of American capitalism were more welcome. But the overwhelming consensus was that we were witnessing a tidal change in business history: the young eating the mature, new media conquering old media, creative destruction, the dawn of the digital economy, or the Internet Revolution.

From my perspective, this wasn't the beginning of the Internet

Revolution, but the end. We were starting to care less about how this technology could augment humanity and more about how it could bolster a flagging stock exchange. The excitement around digital culture, the sexiness of people engaging with one another through media or making new software for free instead of simply watching television all night, was being leveraged to hype a big deal on the old economy's stock exchange. Digital innovations were no longer about changing the world, but keeping the old system firmly in place.

Two months later, the dotcom bubble burst, the NASDAQ crashed, and the very same newspapers that had celebrated the AOL–Time Warner merger were now decrying it as an epic failure on the part of Time Warner's CEO, a former director of the New York Stock Exchange no less, Gerald Levin. People blamed his poor judgment on the trauma of losing his son, and watched, stunned, as $200 billion of shareholder money seemingly evaporated. The Internet Revolution was derided as a Ponzi scheme.

Moreover, the marriage of old and new media did not infuse Time Warner with the disruptive values of internet culture. The resulting company didn't become more adventurous or innovative. Quite the contrary: it became much more traditionally corporate in spirit, obsessed with the bottom line. They fired CNN's Ted Turner—the 1980s' cable television renegade was simply too radical and spontaneous for this supposedly twenty-first-century "new media" company. The promised "synergies" between print and digital media amounted to diluting what was left of revered properties such as *Time* and *Sports Illustrated* with hastily assembled online "extensions." They sold a few banner ads, but ended up mostly giving away online space as a perk to those who advertised in the print publications.

Worse, the new combined company had to justify a tremendous valuation, or market capitalization, to its shareholders. AOL's stock price had been based on wildly speculative guesses about the growth of its dial-up business. But now that AOL's internet subscriber rate had plateaued, the old media side of the new AOL Time Warner was going to have to make up the difference. Somehow. That would be tricky for a traditional publishing company in an era of digital disruption. So instead of finding new revenue sources, the company implemented a tried-and-true method of pleasing shareholders: cutting expenses. Time Warner offered its staff "packages" to resign, slashed research budgets, and even carted out the water coolers. The company sold off productive assets such as Six Flags theme parks and even their cable company, Roadrunner. In order to survive, Time Warner eventually had to spin off AOL entirely, leaving Time Warner back where it was—except without its assets, personnel, cash, or stock value.

So, at least in this case, creative destruction resulted in something more like destructive destruction. It turns out that Steve Case's team had been looking for months for a way to spend his high-priced shares—an exit strategy—while there was still time. Instead of investing in digital innovation, he hired investment bank Salomon Smith Barney to find an acquisition target. This was not any sort of new media strategy. It was plain old buy-low-sell-high horse trading, amplified by the massive bubble of digital finance. As Ted Turner later explained it in characteristic hyperbole, "the Time Warner–AOL merger should pass into history like the Vietnam War and the Iraq and Afghanistan wars. It's one of the biggest disasters that have occurred to our country." And at the time they thought *I* was being pessimistic.

The AOL failure exposed the Ponzi scheme underlying the early internet. The broader dotcom crash that followed erased $5 trillion, or 78 percent, of the NASDAQ index by October 2002. Yes, the markets eventually recovered. But the relationship of technology to capital was forever changed. Money was no longer seen as a way to fund new technologies; new technologies became understood purely as ways of making fast money—as long as you could get out in time.

Venture capitalists burned by the dotcom bust had learned their lesson, too. They would no longer write checks to hackers and wait for results. They felt as though they had internalized what actually mattered about technological development, and should now be in charge. From then on, receiving money from a venture fund meant "pivoting" from whatever a founder may have intended to do with technology toward doing whatever would be most likely to generate "hockey stick" growth curves and 1000x returns for investors on exiting.

The Mindset had arrived.

For example, Google began as a project by two Stanford University students looking for a better way to search the web than Yahoo's top-down classification system. Their bottom-up system would look at the ways websites linked to one another, and use that to determine search rankings. Although the company was wildly successful and profitable simply from putting a few "sponsored links" next to its search results, investors wanted more. Luckily for them, every web search conducted by Google's billions of users also generates a surplus of "collateral data"—whole histories of searches and clicks and other information the company didn't care much about. As Shoshana Zuboff chronicled in her book *Surveillance Capitalism*, instead of simply continuing to deliver search results to users, Google got into the even more profitable business of delivering user data to its real customers—the market researchers seeking to target users and manipulate their behavior.

Likewise, Mark Zuckerberg left college to pursue his (probably borrowed) dream of building an online social network for college students to make friends and get dates. After taking capital from Peter Thiel and others, however, his business model shifted from serving ads to selling data. The longer and more emotionally engaged we are with the platform, the more Facebook can learn about us. Each of our posts, likes, and clicks gets tracked and analyzed in order to encourage still more engagement—often exploiting our vulnerability to sensationalism and amplifying our most impulsive tendencies. The negative impact on our society has yet to be fully measured, and many technology critics have written whole books on the way social media's emphasis on data extraction has alienated us from one another and the factual world. In short, instead of empowering users, his company would enrich investors at his users' expense.

When Facebook's practices of data collection and user manipulation surfaced, I began to give a speech arguing that on these free platforms "we are not the users, we are the product." While catchy, it's not quite true. We are not products of these platforms so much as the labor force. We dutifully read, click, post, and retweet; we become enraged, scandalized, and indignant; and we go on to complain, attack, or cancel. That's work. The beneficiaries are the shareholders. For what Silicon Valley really chose to learn from the AOL debacle is that the true product of any of these companies is the *stock*.

In the new, improved, post-crash version of Silicon Valley, extreme capitalism rules. Digital technology is valued most for its ability to scale a business without needing to hire many human beings, and to provide the earnings or—as is more often the case—the hype required to boost the share price. (Companies that add trendy words like "blockchain" to their names have seen their stocks quadruple.) Following AOL's example of mailing free disks, companies scramble to get subscribers at any cost. A company can lose money for years, as long as its user base is rising—preferably at an exponential rate.

But it's not all abstract. Hockey stick user growth leads to hockey stick stock growth. Then, with the increased capital at their disposal, tech companies build "war chests" with which to lobby for policy changes in the real world. Uber and Doordash spend millions lobbying to be allowed to hire drivers as low-cost independent contractors rather than employees entitled to benefits. Airbnb uses its war chest to fund "independent, host-led local organizations that serve as a forum to connect and gather passionate hosts"—so that they can fight against regulatory pressure from local governments and city councils. By 2017, facing antitrust accusations, Google was outspending every other company lobbying lawmakers in Washington—only to be outspent by Facebook, facing similar charges in 2020.

The capacity that digital companies have for abstracted, exponential growth has allowed them to amass political and economic power unheard of even in the time of the Gilded Age robber barons. Numerous studies have concluded that economic elites now enjoy substantially more impact on government policy, while "citizens and mass-based interest groups have little or no independent influence." As companies lobby to protect their monopolies, small businesses lose the ability to compete. This leads to more bankruptcies and unemployment. Workers have no social safety net because the companies that have rendered them jobless show no profits and pay no taxes. As a last resort, workers turn to gig economy jobs at Doordash, Uber, or Amazon Mechanical Turk, becoming dependent on the platforms that disempowered them in the first place.

The resulting valuations of the larger tech companies—if not their earnings—rival that of many nations. For their part, the people who become billionaires or even centibillionaires off all their stock may start out with good intentions but eventually succumb to The Mindset. Like their companies, they tend to pivot away from helping other people and solely toward increasing their own capital gains. It's as if accumulating wealth in this way has a negative effect on their ability to perceive themselves as members of a society.

Studies have shown that the more power a person has, the less "motor resonance" or mirroring they do of others. Of course, people seeking power may be predisposed to this behavior. But further research has suggested that after people have gained power, they tend to behave like patients with damage to the brain's orbitofrontal lobes. That is, the experience of wealth and power is akin to removing the part of the brain "critical to empathy and socially appropriate behavior." Poorer people are much better than their wealthy counterparts at judging other people's emotions. Their capacity to make "empathetic inferences" based on facial muscle movements is far superior.

Of course, correlation isn't cause, and the specific mechanisms through which the wealthy and powerful lose their ability to feel for others have not been isolated. Capitalism itself, at least as currently practiced in Silicon Valley, certainly supports widespread disregard for the vanquished. Poverty is largely considered the fault of the poor. As NYU business professor Scott Galloway has explained, "we've decided that capitalism means being loving and empathetic to corporations, and Darwinistic and harsh towards individuals." Government readily bailed out banks and businesses in the 2008 recession, and the Covid crisis increased total billionaire wealth from $8.9 to $10.2 trillion in just the first year, despite the pandemic's negative impact on everyone else.

The Mindset encourages a form of "winning" that requires its human and corporate victors to rise above those who have been necessarily left behind. Winning, after all, is by definition a way of setting oneself apart from everyone else. This separateness is the very object of the game, so we shouldn't be surprised that those who reach the top of the pyramid look down on the rest of us. Those who have made it there through questionable means do not want to look back at the devastation they left in their path. They need an exit strategy, and may prefer to imagine a future where they are *forced* to isolate themselves from those they have exploited. Then they won't have to feel any guilt, shame, or fear of retribution.

A Womb with a View

YOU ARE SAFE IN YOUR TECHNO-BUBBLE

What if The Mindset is not just a product of money, but of technology itself?

I remember back around 1990, when psychedelics philosopher Timothy Leary first read Stewart Brand's book *The Media Lab*, about the new digital technology center MIT had created in its architecture department. Tim devoured it, cover to cover, over the course of one long day. Around sunset, just as he was finishing, he threw it across the living room in disgust. "Look at the index," he said, "of all the names, less than 3 percent are women. That'll tell you something." Indeed, while women, and particularly women of color, had been responsible for developing much of the math, code, and languages on which computers depend, they had also been systematically excluded from elite computer science programs and careers.

He went on to explain his core problem with the Media Lab and the digital universe these technology pioneers were envisioning: "They want to recreate the womb." As Leary the psychologist saw it, the boys building our digital future were developing technology to simulate the ideal woman—the one their mothers could never be. Unlike the human mother who failed them, a predictive algorithm could anticipate their every need in advance and deliver it directly, removing every trace of friction and longing. These guys would be able to float, fully fed and serviced, in their virtual bubbles—what the Media Lab called "Artificial Ecology."

Tim's copy of the book, which I've kept to this day, is filled with his angry underlines and marginalia in blue and black ballpoint. The most emphatic comments were scrawled right over the text in red felt tip: *What? Huh? WRONG! NO!!!* In one chapter, he circled Media Lab founder Nicholas Negroponte's original marching orders for Brand to write a book "about the quality of life in an electronic age." Negroponte went on, in a now oft-quoted image, "I was still in my pajamas at ten thirty this morning after I had been doing lab work, through email on my computer, for several hours. Maybe what we're talking about is 'The right to stay in your pajamas.'" The ultimate promise of digital technology as if imagined by a third grader: never having to get dressed or even comb your hair.

Technology would not only play the part of the pampering mother enabling the infantilized user to stay in his pajamas all day, it would also serve as the perfect girlfriend. As Brand explained in the closing passage of the book, "Our machines have to welcome us inside them"—a phrase Leary underlined and ridiculed "Huh! Poor Stewart!" As Leary saw it, MIT's vision for technology was that of a few smart but psychosexually immature white men who wanted all the benefits of being sealed up in their perfectly controlled and responsive environments—without ever having to face the messy, harsh reality of the real world.

For there's the real problem those billionaires identified when we were gaming out their bunker strategies. The people and things we'd be leaving behind are *still out there*. And the more we ask them to service our bubbles, the more oppressed and angry they're going to get, and then the more bubbled we will want to be. No matter how far futurist Ray Kurzweil gets with his artificial intelligence project at Google, we cannot simply rise from the chrysalis of matter as pure consciousness. There's no Dropbox plan that will let us upload mind and soul from the body to the cloud. We are still here on the ground, with the same people and on the same planet we are being encouraged to leave behind. There's no escape from the others.

But digital technologies sure do give us ways of pretending we can.

As Gabe Newell, billionaire founder of the game platform Valve, explained it to *Wired*, "we're way closer to the Matrix than people realize." For Newell, the human body is a mere "meat peripheral" that is resistant to upgrades or repair and "not at all reflective of consumer preference." Virtual reality will give users more "choice" (we'll hear that word a lot) over their perception and experience of the world. His goal is to create a virtual reality with such compelling texture and granularity that we stop measuring simulations by their fidelity to the real world. "The real world will seem flat, colorless, blurry compared to the experiences you'll be able to create in people's brains."

This will be especially valuable as the real world continues to degrade. VR developers even make an economic justice case for throwing us all into a simulation. "It is not possible, on Earth, to give everyone all that they would want," Oculus Rift Chief Technology Officer John Carmack explained on the Joe Rogan podcast. "Not everyone can have Richard Branson's private island." VR is the new solution to climate change—or maybe the ultimate surrender to its inevitability. As resources vanish and economic conditions worsen, technological simulations can fill in where real wealth has disappeared. "The promise of VR is to make the world you wanted."

The Covid pandemic gave us all a lesson in the attraction and limits of such dreams of universal happiness through technologically enhanced bubbling. In most cases, it was the wealthy who bubbled, and the poor who braved the real world to service them. No matter how many mutual aid networks, school committees, protests for racial justice, or fundraising efforts we participated in, many of us privileged enough to do so were still making a less public, internal calculation: how much are we allowed to use our privilege and our technologies to protect ourselves and our families from the rest of the world during this crisis? And, like a devil on our shoulder, our technology was telling us to go it alone. After all, it's called an iPad, not an usPad.

Most of us chose to wrestle with the civic challenges of the moment, such as whether to send kids back to school full-time, hybrid, or remotely. But some of the wealthiest people in my own town chose instead to form private "pods," hire tutors, and offer their kids the kind of customized, elite education they could never have justified otherwise. "Yes, we are in a pandemic," one pod education provider explained to a *New York Times* reporter covering the phenomenon. "But when it comes to education, we also feel some good may even come out of this."

The speed and completeness with which so many embraced home delivery, Zoom meetings, and streaming *Hamilton* on DisneyPlus made it hard to tell: was it really Covid inspiring this drive toward screens over contact, remote learning pods over public education, relegation of undesirable tasks to the poor, and widespread retreat of the privileged to vacation homes protected by doorbell surveillance cameras—or was the pandemic simply helping to justify a trend already well in progress? Were we panicking, or was the billionaire prepper ethos trickling down to the middle class? Or both?

Not without a touch of guilt, many of those who could afford high-tech solutions to life under lockdown learned to embrace their predicament. "I finally caved and got the Oculus," one of my best friends messaged me just two weeks into the lockdown in our area. "Considering how little is available to do out in the real world, this is gonna be a game changer." Between VR, Amazon, FreshDirect, Netflix, Zoom, and a sustainable income doing web services and crypto trading, he was going to ride out the pandemic in style.

The problem is, the very technologies we use to connect under these conditions also undermine our empathy for those outside our Covid bubbles. We establish rapport with other people through subtle social cues evolved over centuries to promote partner bonding and group sharing. When we engage with others in real life, we can see if their pupils are growing larger to take us in, if their breathing rate is syncing up with ours in empathy, or if their faces are flushing with passion. This, in turn, fires the mirror neurons in our brains, stimulating a positive feedback loop and releasing oxytocin—the bonding hormone—straight into our bloodstreams.

On a Zoom call, much less a text message or Twitter comment, we can't feel these subconsciously detected cues. Someone may say they agree with us, but we can't confirm this assertion with our bodies. The mirror neurons don't fire, the oxytocin doesn't flow, and we're left in a state of cognitive dissonance: they said they agree with me, but it doesn't *feel* like it. Our bodies don't know to blame this on the media environment. Instead, we blame it on the other person. They are not to be trusted.

This sense of distrust and alienation then feeds back into the way we write our business plans and build our technologies. *Those are not people.* They're just users or gig workers on the other side of the screen. That makes them easier to surveil, exploit, dominate, ignore, and leave behind. The Mindset's logic becomes self-reinforcing.

Unlike the billionaires, many of us didn't like the moral compromises we made during the pandemic, but felt little freedom to choose otherwise. Sure, I donated my government relief check to the local food pantry, and sent a significant portion of my stable income to friends who could no longer meet their basic expenses. But I also went and spent five hundred bucks on a big plastic pool for my daughter and our neighbors' kids to use as the basis for a makeshift private summer camp. And I saw similar inflatable blue bubbles throughout town, all eventually destined for landfill.

Of course, the pool wouldn't have arrived were it not for legions of Amazon workers behind the scenes, getting infected in warehouses or risking their health driving delivery trucks. After learning of the way Amazon avoids taxes, engages in anti-competitive practices, and abuses labor, many progressives once swore off the platform. But there we were, reluctantly re-upping our Prime delivery memberships to get the cables, webcams, and Bluetooth headsets we needed to attend the Zoom meetings that came to constitute our own work. Others reactivated their dormant Facebook and WhatsApp accounts to connect with friends, all sharing highly curated depictions of their newfound appreciation for nature and homemade bread. Covid-specific desperation turned otherwise dubious startups like Clubhouse (an audio chat platform) and Onlyfans (for webcam sex workers) into overnight sensations. As these compromised platforms replaced our social lives, many of us were lulled further into digital isolation—being rewarded the more we accepted the logic of the fully wired home, cut off from the rest of the world.

The more blatantly people embraced their sequestered life-styles, the better many of them appeared to do. It was as if they were advertising the perks that a digital existence had been offering us all along. Under the pandemic, more people opened online trading and crypto accounts than ever before, and got rich off these increasingly hyped and crowded video game versions of the marketplace. On YouTube, Clubhouse, and Twitter, millennial crypto traders shared their winning strategies along with photos and videos of the Teslas and overpriced NFTs they purchased with their bounty. Likewise, groups of social media celebrities moved into luxurious "hype houses" in Los Angeles and Hawaii, where they livestreamed their lifestyles, exercise routines, and sex advice—as well as the products of their sponsors—to their millions of followers. Things may be bad out there, they seemed to be saying, but if you buy into the digital bubble, life *can* be a cabaret.

The digital platforms supporting this lifestyle were rewarded under Covid, too. Shares of Zoom went up more than 700 percent during the first ten months of the pandemic. Amazon founder Jeff Bezos's fortune rose $86 billion in about the same period. While airlines, hotels, and brick-and-mortar businesses struggled or went under, the combined revenue of the five biggest U.S. tech companies—Apple, Microsoft, Amazon, Alphabet, and Meta—grew by a fifth, to $1.1 trillion, and their combined market capitalization rose 50 percent to over $8 trillion by the end of 2020. Netflix's share price rose more than 60 percent in the first months of the pandemic. Each new strain of the virus led to another stock boost for the same companies.

These profits enabled the executives of such companies to pay upwards of $70,000 a month for wifi-enabled warm-weather resorts in Hawaii, Costa Rica, and Belize. The population of wealthy executives seeking to telecommute from paradise grew so large that new agencies and facilities emerged to cater to them. And, unlike a fortified billionaire's bunker, most of these deluxe rental communities gave their residents a chance to network with others just like them. "Many creatives, startups, and techies are realizing they can meet interesting investors in places like Oaxaca or San Miguel de Allende," one travel adviser explained to *Bloomberg* magazine.

The dependably wealth-apologist *New York Times* was busy running non-ironic photo spreads of families "retreating" to their summer homes—second residences worth well more than most of our primary ones—and stories about their successes working remotely from the beach, or retrofitting extra bedrooms as offices. "It's been great here," one venture fund founder explained. "If I didn't know there was absolute chaos in the world . . . I could do this forever."

That chaos in the world was real. While the wealthy retreated, the poor were clobbered. Each 1 percent increase in a county's income inequality was associated with a 2 percent increase of Covid infection and a 3 percent rise in related deaths. By nearly every metric, the poorer a region or country, the more Covid and death. Likewise, the people processing pork and beef suffered over 100 percent greater transmission rates than the people to whom all that meat was delivered. Anywhere we look, the sad plight of those left behind in the chaos remains the same.

But what if we don't have to know about the chaos out there in the real world? That's the true promise of digital technology. We can *choose* which cable news, Twitter feeds, and YouTube channels to stream—the ones that acknowledge the virus and its impacts, or the ones that don't. If we have the money, we can simply filter out what we don't want to see—or, better, watch just enough to justify our decision to cover our eyes. Maybe that's why the most popular TV show streamed during the pandemic—which swiftly became the most popular show in Netflix's history—was *Squid Game*, a South Korean allegory of the competitive cruelty of capitalism, in which people destroyed by the marketplace voluntarily enter a lethal gaming world for the entertainment of a handful of billionaires. On some level, those of us living mostly online identified with both the poor souls who ventured into the game and the elite who were watching them compete from a distance.

The point is not to condemn those who succumbed to fear and spent everything they could on their own family's safety. It's that the Covid pandemic offered a ghastly trial run for a fully digitally immersive future. It painfully, even shamefully revealed how the insulation equation is latent in all of us, as well as how it was exacerbated by the technologies that mediated the entire experience.

More tellingly, Covid gave many of us the excuse to finally submit to one of the dominant elements of The Mindset's ethos—the one those billionaire preppers live by—which is to design one's personal reality so meticulously that existential threats are simply removed from the equation. The leap from a Fitbit tracking one's heart rate to an annual full-body cancer scan, or from a doorbell surveillance camera to a network of autonomous robot sentries, is really just a matter of money. We are all, to some extent, in this same game.

The best technology can really offer is an illusion of insulation. Whatever may be befalling everyone else—whether it's an isolated virus or systemwide climate catastrophe—technology can make us feel protected. The final irony of using digital technology to titrate one's exposure to the outside world is that it perpetuates the concomitant illusion that we're all somehow distinct from one another when we're *not* plugged in. In other words, no matter how deeply immersed we are in the video game of online life, the real world of viruses, poverty, terrorism, climate change, and other horrors persists. We simply become less able to empathize with it, mitigate its effects, or prepare for its eventual encroachment on our lives. So much the better for the algorithms streaming us the picture of the world we want to see, uncorrupted by imagery of what's really happening out there. (And if it does come through, just swipe left and the algorithms will know never to interrupt your dream state with such real news again.)

Like a hiding toddler who thinks holding their hands over their eyes can prevent them from being seen, those who would rely on digital technology to mediate the world are in for a surprise. The Oculus VR headset has a "guardian boundary" that keeps users from crashing into walls while they're playing in virtual worlds. But climate, poverty, disease, and famine don't respect the safe play space defined by our user preferences. No matter what technology we're using, none of us can climb back into the womb.

The Dumbwaiter Effect

OUT OF SIGHT, OUT OF MIND

We can't blame capitalism for all of technology's ills, nor can we blame big tech for the devastating excesses and blind spots of business. But neither business nor digital technology could have wrought the havoc of this moment on their own. Rather, the two have formed a mutually reinforcing feedback loop that encourages entrepreneurs to envision a future ruled by private sector technologies that work to make our problems invisible, even if they fail to solve them.

The would-be architects of the human future treat the civic sector as antagonistic to their grand designs. They believe they can do it better. Unencumbered by any consideration for the impact of their projects on the rest of us, they can no doubt build spectacular things faster and more profitably than any government. But this requires sweeping a lot of stuff under the rug—like the people and places where their systems actually operate.

For example, usurping the role of municipalities in planning mass transit, Uber commissioned eight leading architecture firms to develop design proposals for "skyports" where future users of the ridesharing app could board and disembark from as-yet-uninvented urban air taxis. In spite of their lip service to social and environmental impact, the proposals convey a future like the one depicted in Fritz Lang's *Metropolis*, where the wealthy fly from point to point in a city in the sky—while the workers supporting this lifestyle toil on the ground below.

Although architecture firms have a tendency to justify their designs with lofty language, some of the descriptions offered for Uber's skyports appear to be intentionally abstract. "The integration of all Uber brands substantiates first and last mile travel as major support elements to the Uber Air component that revolutionizes urban mobility," explains one of the proposals. "The Mobility Hub is not a thing, but rather a place of dynamic energy and integrated connectivity that celebrates the spirit of flight and the freedom to quickly access the important places in one's life."

Down here in the world of "things" that Uber seeks to transcend, cars are increasingly becoming sleeping quarters for homeless families. More than one-third of schoolchildren in the East Palo Alto school district, the town neighboring Facebook's headquarters, are homeless. Envisioning a different sort of mobility hub, the superintendent has proposed outfitting the school parking lot to accommodate RVs and those living in cars.

As if allergic to the input of municipal leaders, community representatives, or advocates for the poor, those charging themselves with building civilization 2.0 act as if their tech capabilities combined with their private sector successes entitle them to program a new world from scratch, for profit. If Jeff Bezos already controls Amazon Web Services—the infrastructure through which over a third of our networked interaction already takes place—why shouldn't he be the one to build the space program through which humanity migrates to its next home? If Elon Musk could become the richest person in the world by transforming the automotive industry with his Tesla cars, shouldn't he be entitled to realize his dream of colonizing Mars with giant domes?

When the tech fetishist's childlike hope for a digital womb combines with the billionaire's faith in a winner-takes-all competitive marketplace, look out. It results in a brand of activist futurism that sees the present—our reality, including us—as an impediment to their vision of what could and what should be. It's an even darker corollary to the insulation equation that simply seeks to distance the wealthy from the collateral effects of tech development. Here, technology is called back into service as the very means through which all that human suffering is then hidden from sight.

After all, space stations and Mars colonies for the lucky few don't get built without legions of workers left behind on the dying home planet. It's a heck of a lot easier and less stressful to stay focused on 3D models of interplanetary spacecraft or animated renderings of interlinking seastead communities than to consider the lives of the people who would have to make the Doordash deliveries to such places. The only substantive difference between today's reality and the one in their high-tech fantasies is the absence—or at least the invisibility—of the working poor.

On some level, those who intend to create a future based on The Mindset understand the harm they must do in order to maintain their privilege. Their business models almost universally depend on exploiting both consumers and the labor who serve them. While these companies addict our tweens to social media and our crops to glyphosate, they also send slave labor into caves to mine for rare earth metals and out onto toxic waste dumps to forage for "renewables." People who are exploited this way are bound to get angry, or even dangerous.

That's why the first thing wealthy preppers worry about when fantasizing doomsday bunkers is how to maintain the allegiance of the mercenaries protecting them. A revolt of the masses is not a hypothetical fear. Even without a revolution, the suffering masses are hard to look at and think about. No matter how good they get at repressing it, tech bros can't help but experience twitches of empathy when witnessing the suffering of others. Digital technology provides the perfect window for keeping an eye on the oppressed without allowing those compassionate instincts to kick in. Under the guise of increased connectivity, social media helps engender an entirely less compelling and experiential form of connection. It's a welcome feeling of alienation for those who want to be able to bark orders via text message without looking into the eyes of the human being on the other side.

Let's call this relationship to technology "the dumbwaiter effect," after Thomas Jefferson's ingenious food delivery system. We were taught in school that Jefferson invented the tiny manual elevator so the enslaved at Monticello wouldn't have to trudge up the stairs with plates of food. They could put a tray inside the mechanical chamber and then hoist it up with a pulley. The upstairs attendants just open a small door, and voilà: dinner is served. But the dumbwaiter had nothing to do with saving anyone a trip up the stairs. The food was already being carried through underground tunnels and up multiple stairways. The real purpose was to spare Jefferson's dinner guests the sight of the enslaved servants huffing and puffing. The food simply appeared. No observable human suffering.

Too many of today's technological processes stem from this same urge to distance consumers from the reality of labor. In the final stage of cell phone assembly, for example, workers wipe down each unit with a toxic solvent that removes their own fingerprints from the devices. The chemical leads to miscarriages, cancers, and shortened lifespan. The benefit, of course, is that it removes all traces of human involvement. Consumers open the box (perhaps making a video of the "unboxing" ritual) to reveal a piece of electronics that may as well have been teleported from a factory in another dimension. There are no human fingerprints to remind us of the factory conditions in China where it was actually made. In order to remove evidence of suffering laborers, tech companies poison them further.

Some of Amazon's most clever innovations exist entirely to shield Prime members from the reality of working for the company. Its platform and apps are designed to be addictively fast and self-contained—push-button access to stuff that can be left at the front door without any human contact. "Touchless." The delivery people don't even ring the bell; a photo of the package on the stoop automagically arrives in your inbox and Alexa issues a friendly alert. We don't have to confront the poor soul who drives the truck, much less the ones scurrying around between the robots in the warehouse.

Or take Amazon's new custom T-shirt fabrication service, MadeForYou—likely just a trailhead for a generation of future custom products. Consumers use a phone app to take pictures and measurements of themselves, which are then processed into a "virtual body double." Robots then calculate, cut, and sew the custom shirt, and ship it directly to you. It's the ultimate in human personalization, brought to you by machine. Or, as Amazon put it in their promo video, "create the perfect tee, custom-made in size YOU." It's not only the pinnacle of the culture of the autonomous individual—you get your own name on the label—but also a demonstration of the power of automation. Why get shirts made by poorly treated Chinese or even American assembly-line workers when you can get them made by machines that know what you really want?

The illusion here is that the technology is doing all the work. Just talk into your Alexa and Jeff Bezos's army of automated robots will snap into action to make your clothes. His business objective is to own the interface between us and his robot workforce. MadeForYou is his proof of concept for the future of everything from automobile manufacturing to military drones. But robots are not truly doing all the labor. They may be sewing the shirts together, but human beings are still picking the cotton, mining the raw materials for the robots, and burying the obsolete ones in landfill. The externalities are still there—from labor to pollution to resource wars. The robots aren't replacing the human toll so much as hiding it. The dumbwaiter effect.

The layers of obfuscation afforded by digital technology allow for a whole new level of externalized harm. Whether using drones to bomb a social gathering or algorithms to calculate jail sentences, the technology separates the human actor from the human cost. The problems inherent in violent, extractive, growth-based capitalism are not mitigated, but rendered as invisible as the fingerprints of the poisoned workers assembling our cell phones.

Most of us learn to look the other way, at least most of the time. We drive cars, stream video, invest in crypto, and purchase cheap electronics, thankful—on some level—that our highly technologized society puts a few layers of obfuscation between our moment-to-moment choices and their real impacts on the world. Still, how can otherwise sane, occasionally rational, and often empathetic people live this way? Most of us are either ignorant of the harm we cause or so enmeshed in systems beyond our control that we just do the best we can and try not to think too much about it.

Some of the wealthiest and most powerful among us, however, have come to accept the insulation equation as a fundamental principle of our world. Amazingly, it works well enough to make them billionaires in the process, confirming the validity of their convictions to themselves and to their growing legions of acolytes. They become our society's heroes. Cherry-picking compatible ideas from science, economics, and philosophy, they have assembled a mindset that actually encourages them to build a highly technologized society capable of supporting denial at scale.

Selfish Genes

SCIENTISM OVER MORALITY

In what must have been a practiced routine, Richard Dawkins leaned over and began folding a piece of paper on the glass coffee table. The famed evolutionary biologist was the guest of honor at a dinner party in the Central Park West apartment of literary agent and science world luminary John Brockman. It was still just the twentieth century, and most people didn't yet know what memes were.

"Have you ever seen one of these?" Dawkins asked as he held up his completed origami fortune teller for the dozen or so members of the New York intelligentsia who were gathered around the sofa. Of course, we all remembered making and playing with what my childhood friends called a cootie catcher. "This is a meme."

The fold-up paper fortune teller varied very little over time, he explained. That's because it is not simply an object, but a set of instructions—a way of folding paper into a specific configuration. Kids learn how to do it, and then show their friends. Those instructions pass from person to person, both geographically and through history. The people, he explained, are just carriers for the meme. We execute its instructions, like an organism following its genetic code or a computer running some lines of Javascript: fold the corners together like this, then fold sideways like that.

It didn't make sense to me. My own book on viral media was based on very different observations about our relationship to memes. For me, memes were just the code within media viruses, and better understood as "hidden agendas in popular culture." The Rodney King tape may have contained powerful memes about police brutality and race relations, I explained, but the reason the tape became a phenomenon was our readiness as human beings. We had repressed this subject for too long, so we were triggered into a national conversation—and conflict—by the images that forced the issue into the light of day. The memes aren't running us like software runs a computer; we use the memes the same way we use language or our bodies to express ourselves and enact change. We are conscious actors, not passive machines running code.

Dawkins dismissed my argument as "wishful thinking." I was a bit too shy, and too honored simply to be at the same party with him and other science celebrities, to say much more than I already had. I left Dawkins to other admirers and circulated through the room. Half an hour or so later, though, I heard some raised voices coming from the couch.

Naomi Wolf, the feminist author of *The Beauty Myth* and one of the only women at the party who wasn't there as a "plus one," was refusing to accept Dawkins's highly mechanistic picture of the world. "You mean there's absolutely nothing else going on here?" she asked rhetorically. Wolf felt that Dawkins's model of human functioning was unnecessarily reductive and left no space for the underlying mysteries of human perception, experience, and will—not to mention the possibility of a soul or God.

Dawkins was arguing that we live in a universe that can be completely explained by simple scientific principles. It's all purely empirical; we are just clusters of organic matter. Any other frame of reference amounts to superstition, religion, or "delusion." He dared Wolf to come up with evidence of anything that can't be explained by science—anything outside its materialist principles. But no matter what she offered—love, hope, intuition, faith—Dawkins said these were all just emotional states or mental processes designed in one way or another to keep our genes replicating. There was nothing "out there," or even *in* there.

I took a stab at it. I suggested that the universe "leans" in a certain way. "Evolution is not just random selection," I offered, "but life groping toward something. Complexity. Consciousness. Compassion. We're not just driven by genes. The earliest humans shared food with each other, even when there was no personal benefit. Human evolution is not best characterized by competition. It's a story of collaboration."

He and some of the other men laughed. He said I was misinterpreting "reciprocal altruism"—the necessary, but very provisional and temporary ways we cooperate in order to ensure the survival of our genes. Any empathy or urge to share with others is a stimulus generated by our DNA for its own selfish ends. Dawkins and the other scientists explained that we humans are just vehicles for our genes. Everything we think or imagine or want or hope for is in service to them. Our consciousness is itself an illusion enacted by our genes. And since our genes only act selfishly, we humans act entirely selfishly, ourselves.

They kept running logical rings around me. I realized that I needed to stop trying to defend alternative models of reality with the evidentiary rules of orthodox scientism. Centuries of philosophers have shown in multiple ways that scientism—the refusal to consider anything without evidence—is severely limited, I offered. It's great for building bridges and flying airplanes. But the way we ascribe meaning to things is not based on evidence. It's a socially constructed system, built over millennia. And while it may not help us build airplanes, it can help us decide whether we want to have another airplane in our world, and whether to use it for travel or warfare. Only a community of people making meaning together can ground objectivity in any real sense of purpose. Meaning is the way human beings develop a more robust sense of justice—of right, and wrong.

Dawkins rolled his eyes, and explained that religions have led to more wars than science. I told him I wasn't religious, and I don't have to believe in God in order to believe in a just universe. Living as humans in a meaningful universe simply means existing in a reality that has certain rules to it, I suggested. A potentially moral universe. And this goes along with having a sense of right and wrong—an ethical sensibility that goes beyond mere survival.

He had spent more than enough time on this tangent, and finally dismissed my argument as "moralist." People laughed. Like that's a bad thing. A silly, unarguable position.

It's not just that Dawkins and I had different understandings of the world. It's that he thought he and the other scientists were the only ones seeing things as they *really were*, while Wolf and I were merely interpreting reality through our systems of meaning and morals. He couldn't acknowledge that his own commitment to scientism is based on something passional—something more like *faith* in an empiricist universe. In other words, his insistence on living in an evidence-only universe isn't based on evidence at all.

It's an assumption. It's part of a system of meaning, developed by a community of people over time. It just happens to be a meaning system that ignores meaning itself. Worse, by rejecting the validity of any other meaning system, it is prone to instilling in its adherents a sense of superiority over others. Those who strive for meaning are mere "moralists."

So, Stephen Pinker argues for a computational theory of mind—one where the brain is mere hardware playing various programs in utterly predictable ways. Daniel Dennett likens religious belief to a "parasitic worm (a lancet fluke) invading an ant's brain." Under scientism, humans are just robots running programs—either the ones dictated by our genes, or destructive pathogens like spirituality. However, by refusing to understand how meaning-making is a subtle community project related to the ways we live together, this orthodox scientism denies any scheme of things where human agency—hand in hand with moral responsibility—plays a role. We are at the mercy of our programs, our genes. So we may as well do whatever the heck we want, especially if it spreads our genes around.

A little more than twenty years later, I recalled my encounter with these thinkers as I read the magazine stories about the fall of serial underage sex trafficker Jeffrey Epstein, and the various scientists he funded in his efforts to revive the practice of eugenics and seed hundreds of women with his own sperm. Photos surfaced of Epstein and several of his young escorts together with many of the very scientists who reject citizenship in a moral universe as laughably deluded. Another showed Daniel Dennett, Stephen Pinker, and Dawkins himself flying to the TED conference on Epstein's private jet, the unambiguously nicknamed Lolita Express.

While anyone can end up in the wrong place at the wrong time, the scientists with whom Epstein chose to hobnob weren't selected at random. Their decidedly scientistic approach to human development, interpreted through Epstein's sociopathic lens, dovetailed ever-so conveniently with his master plans for the human race. Epstein was truly the model, self-sovereign, transhumanist billionaire prepper. He owned a private island (where he made his own laws) and several retreats—including a ranch in New Mexico where he planned to house and impregnate twenty women at a time. He gave millions to scientists he considered too maverick for the politically correct sensibilities of modern funders and institutions—researchers he believed could help him dominate the human gene pool, avoid death, or, if necessary, freeze his head and penis for future reanimation.

This is not science or even scientism; it's just The Mindset. The scientific model is a method of collaborative inquiry, and perhaps humankind's greatest single achievement. It has supported not only those who seek to understand the nature of our reality, but also those who want to help us live better, healthier lives, distribute food, energy, and resources more effectively, document our impact on the environment, and understand our place in the universe. If anything, we suffer from too little faith in good science, not too much.

Yet, forcibly removed from the greater contexts of meaning and morality, science easily falls into the service of domination and control, providing justification for the most alienated features of The Mindset. After all, the original premise of empirical science, as articulated by (or at least attributed to) its forefather Francis Bacon, was hardly value-free; it was based on the subordination of nature and women. As he reportedly explained to his seventeenth-century benefactor, King James, nature would have to be forcibly penetrated to yield her secrets. "I am come in very truth, leading you to nature with all her children to bind her to your service and make her your slave," Bacon explained. "Nature must be taken by the forelock . . . lay hold of her and capture her . . . conquer and subdue her, to shake her to her foundations." Such statements suggest that nature should be subjected to the same sorts of interrogative torture as women on trial for witchcraft. For nature "exhibits herself more clearly under the trials and vexations [of mechanical devices] than when left to herself."

Nature was scary, dark, and female—a boundless and all-encompassing space of mystery. Empirical science could capture and tame this beastly force, quantifying its properties and rendering it inert. Anything that couldn't be observed and quantified did not exist. This made for a vastly oversimplified but comforting picture of the world as driven by entirely measurable and predictable phenomena.

From the beginning, the promotion of empirical science depended on the subjugation of women as witches. Of course, this activity was not internally consistent with the logic of scientism. If witchcraft were mere superstition, then the women practicing it would not have the powers ascribed to them in the *Malleus maleficarum* (the Hammer of Witches), which was used as a field guide in their capture and identification. But Bacon and the Royal Society he inspired were in a difficult position.

First off, women healers, using a very different language for the properties of herbs and curative methods, were still proving themselves entirely more effective at treating medical conditions than the men of science with their leeches and bloodletting. Second, the Royal Society and its scientists were themselves being accused of "atheistical" materialism by the Church. By acknowledging the power of black magic and joining the witch-burning frenzy, men of science distanced themselves from their own god-less, materialist picture of the world, while also eliminating their greatest competition.

In the process, they robbed themselves of experiential approaches to science and several thousand years of retained knowledge from those practices—including several centuries of trial-and-error experimentation with herbs, agriculture, and weather prediction. They also internalized a chauvinism and disregard for alternative approaches to making meaning that have limited scientific discovery and discussion to this day. Early empirical scientists sought to contain the forces of nature by quantifying them. Everything had to be measured in terms of mass, heat, force or some other metric. The quantification of our world contained and controlled everything that mattered, while ignoring the pesky, undefined aspects of reality that men of science didn't want to mess with—especially emotions, meaning, and ethics. Empiricists were great on the whats and wheres, and not so great on the whys.

By divorcing itself from meaning systems (especially the ones from which it emerged), science made itself particularly vulnerable to forces that sought to leverage it for domination and extraction. Renaissance-era science, which categorized reality from above, was compatible with the top-down, one-size-fits-all approach of the industrialism that followed. Monarchs and their chartered monopolies could pick which variable they wanted to maximize—speed, output, profit, distance, carnage—and scientists could develop machines and processes to match.

Empirical science also conveniently separated causes from effects. Things acted upon each other, but were not understood to be in dynamic relationships. Something or someone was either a subject or an object, solute or solvent, predator or prey, man or woman, lord or peasant, master or slave. The further that the powerful in a scientistic society could separate cause from effect, the less they had to look at what they were doing to whom.

Thanks to early scientists' objectified, quantifying, transactional biases, science and the technologies it spawned became the fast friends of colonial capitalists who were looking for ways of putting a monetary value on everything. Scientific engineers developed the technologies that armed the gunships, insulated the powerful from the impacts of their actions, and rationalized an ethos of extraction. Worst of all, though, this repressive scientism ended up perpetuating a legacy of domination and control among future generations of science's Royal Society.

It's how you get a biologist as brilliant as Richard Dawkins reducing the mysterious phenomenon of human consciousness to nothing more than a movie projected for us by our genes. But my own experience with computer code taught me the opposite of what Dawkins concluded about human beings from his study of genes. Where Dawkins saw human beings as "survival machines—robot vehicles blindly programmed to preserve the selfish molecules known as genes," I saw sharing, collective awareness, and a new renaissance of collective creativity.

Dawkins's model of human-as-hardware, and those of intellectually compatible colleagues like Daniel Dennett and Stephen Pinker, won out in Silicon Valley. Their views were entirely more compatible with business models that depended on manipulating human beings instead of empowering them—exploiting them for profit rather than giving them opportunities for collective creativity. If people were really just passively responding to lines of genetic or cultural code, then why not be the ones writing that code and capitalizing off it?

Over the next decade, an increasing number of seminars and retreats organized by Brockman's Epstein-funded Edge Foundation were dedicated to behavioral economics—the study of why people make the sorts of financial decisions they do. Behavioral economics is really just a euphemism for marketing psychology, from an even more programmatic perspective. So it's not surprising that attendees of these master classes in behavioral economics included Amazon founder Jeff Bezos, Google's Larry Page, Microsoft CTO Nathan Myhrvold, and Tesla's Elon Musk, soaking up actionable wisdom from the authors of books with titles like *Nudge* and *Misbehaving*.

Behavioral economists are most interested in pinning down the ways in which people diverge from acting in what should be their selfish economic interests, for these are the "exploits" or levers through which consumers and investors can be manipulated to spend money irrationally. For example, thanks to the irrational process of "mental accounting," people tend to consider their money as existing in separate buckets—some of which feel worth more than others. Loyalty schemes such as credit card points and airline miles encourage mental accounting, leading people to spend more of those assets than they would otherwise. "Anchoring bias" refers to our tendency to rely on the first information we hear. Marketers exploit this by telling us an "original price" before the "sale price," making consumers believe they are getting a bargain. Behavioral economics is just another form of binding nature—in this case, human nature—to one's service by treating people like programmable machines.

Just as science's seventeenth-century champions self-interestedly gave lip service to the values of the church, those promoting applied science of this kind are selling out to the values of the market. They are helping formulate a picture of the world that gives cover to those who would exploit other human beings: people aren't really alive or aware—they're just behaving in service to their genetic programs. They are primates subject to biases and blind spots. It's a sociopathic perspective that makes science valuable to the billionaire class because it helps justify their most shameful behaviors—from trafficking young women to exploiting an entire underclass of workers and consumers.

These fans of science are perpetuating an unacknowledged value system inherited from the forefathers of the scientistic tradition; it fully informs their actions and assumptions, even though they believe themselves to be nothing but evidence-based rationalists. In reality, they're the ones just following a program: The Mindset.

Pedal to the Metal

DEHUMANIZE, DOMINATE, AND EXTRACT

My friend Bernie and I studied theater directing together as master's students at California Institute of the Arts. It was a great school, with all sorts of opportunities for interdisciplinary collaboration. It's where Tim Burton (director of *Batman* and *Beetlejuice*) learned animation and K. P. H. Notoprojo (the world's most renowned Javanese gamelan performer) taught the famed "monkey chant" to several hundred at a time, while Paul Reubens (PeeWee Herman) roller-skated down the corridors in a tutu.

The theater department, however, was decidedly traditional. Following the conservatory model, they taught the classical approach to drama: crisis, climax, resolution. Every play, every scene, every moment, contained that same shape, propelling the action ever forward toward the protagonist's singular goal or destiny. Every beat had an inciting event to trigger the action. The action escalated to a peak. Then the protagonist made a choice— kill, marry, escape, commit suicide—and the story resolved. Crisis, climax, release. Aristotle described it, Shakespeare perfected it, and Hollywood turned it into a formula.

Bernie and I didn't see theater that way. He was trained in dance and mask performance. He saw theater as an open-ended, highly improvisatory, rhythmic exploration. Why should a performance have to be rational or explicable in terms of human motivations and goals? Do people really know what they want all the time? For my part, I was raised on the experimental theater of the seventies and eighties, where the line between performer and audience was blurred beyond recognition. The story of the play was just an excuse to put actors and observers in a room together, with an imaginary divide between them. I studied "happenings" and fluxus, which were less like theater scripts than they were rule sets from which situations emerged. I didn't like performances that went in straight lines, with beginnings, middles, and ends, because life wasn't really like that.

"This theater we're doing isn't art," I remember complaining to Bernie one night after class.

"It isn't even life," he agreed. "Since when have you experienced yourself as a protagonist driving toward a goal? Shit just happens. Then it stops. Or maybe starts again."

"They call this realism, but it's the opposite," I finally added. "They're imposing a narrative. A sense of inevitability. To confirm the established order. Rising action, climax, sleep. Like a male orgasm curve. Drive toward the goal, then roll over and pass out."

"It's cultural propaganda," Bernie concluded. "Create a problem in the first act, and then solve it the last. Whatever your solution— war, love, god, honor—that's the value the audience learns."

On top of all that, theater was expensive. I remember attending a production of Brecht's *Threepenny Opera* where the cheapest ticket was seventy bucks. The play intentionally spoofs and undermines some of the conventions of theater and, in particular, the romantic extremes and happy endings of Wagnerian opera. In Brecht, the people with big plans for the future are usually the villains. They ignore current pain and anguish as they steamroll toward their visions of domination. Their ends always justify their means. Brecht's plays were meant to expose and undermine that forward drive toward conquest and conclusion.

During the intermission, I heard a woman complaining, "But where is it *going*?" Audiences had come to expect forward motion toward some goal. That's what they were paying for. Life certainly didn't offer such satisfaction. Entertainment was supposed to.

That was the day I decided to leave the theater for the internet. Interactivity would change all this, I told myself. And to a point, it did. The Web, digital platforms, and hypertext stories offered multiple pathways for users to follow or even forge. There was no longer one goal to achieve; we were free to choose our own adventures. Even in games with clear-cut goals, from Super Mario to World of Warcraft, players could find great satisfaction ignoring the official story and wandering around in the world of the game instead.

But while the activities taking place on these platforms may not have been characterized by a single-minded drive toward conquest, the businesses underlying them sure were. In the 1980s, clever people wanting to strike it rich out west would write high-concept screenplays. By the 1990s, those same clever people were writing tech business plans with pretty much the same mechanics. A big new idea will "disrupt" the status quo, take out the competition, grow the market to its full potential, and then—at the peak—execute its climactic "exit strategy" as a sale or IPO. Beginning, middle, and glorious end. Narratives of triumph, expressed in Return On Investment.

The rhetoric of Silicon Valley—whether in the pitch decks of young developers, the talks by TED speakers, or the Joe Rogan interviews with tech billionaires—always bears the same hallmarks as these business plans. Progress. The future. Optimism. Transformation. Winning. But usually these are just euphemisms for conquest, colonization, domination, and extraction. They describe ends-justify-the-means campaigns to change the landscape and achieve a monopoly.

Critics of the Western, linear drive toward progress too often rely on a romanticized picture of the distant past as somehow free from cruelty and violence. They offer accounts of peaceful interactions between primates such as bonobo apes as evidence of our cooperative ancestry. In this view, sustainably living indigenous peoples were infected by the violence of expansionist empires, and now that we know this, we've got to make like Joni Mitchell and "get ourselves back to the garden." On the other hand, proponents of rapid development offer a utopian vision of the future where technology and the market liberate humanity from the darkness of its competitive nature. We descend not just from violent apes but from violently competing bacteria. Civilization, markets, and technology give us ways to channel that innate competitiveness toward better outcomes for all.

Both narratives are steeped in the ideology of progress and the mythology of some fundamentally different place to which we're all going—or to which at least some lucky few of us are going. The tech bros and their most ardent antagonists fall into the same mental trap. Whether corporate or counterculture, these conquest narratives all follow what we might call heroic journeys or New

Testament architectures. Struggle, progress, climactic apocalypse, and then salvation for those with the right belief, psychedelic experience, computer processor, or selfish genes. Those lucky winners get the final, transformative climax to infinite wealth, Mars, eternity on a chip, or Christ consciousness. They finally arrive somewhere, and the story ends.

It's not that this narrative shape was born at some moment in the history of Western civilization and then took over the world. Rather, the drive toward conquest is a human potential that gets activated and amplified by certain kinds of discoveries and innovations—and then supports those discoveries and innovations in return. Cultural historian Riane Eisler traces one instance of the "dominator model" back to the beginning of the Iron Age and the mastery of metallurgy by the Kurgans of early Europe. "The power to dominate and destroy through the sharp blade gradually supplants the view of power as the capacity to support and nurture life," she explains. "Men with the greatest power to destroy—the physically strongest, most insensitive, most brutal—rise to the top, as everywhere the social structure becomes more hierarchic and authoritarian."

Everything changed to support the dominator mindset, until people born into this new culture assumed its rules and values as the given circumstances of nature. The Kurgan invaders were so successful, in fact, that they were eventually idealized by Nietzsche and Hitler as the only original, pure European race. It's a dynamic we'll see repeated a lot: a new technology emerges, someone copies the idea then uses it to colonize a market or culture, break it down into smaller, alienated groups, and extract its resources and labor—all while supplanting existing values with an ideology of competition and domination. In fact, it's the very basis of what we now call capitalism.

Initially the market economy, introduced just after the Crusades in the late Middle Ages, benefited the former peasants of feudalism. This was a lateral, peer-to-peer economy, not a hierarchical one. Local farmers and bakers didn't generally aspire to be "rich" so much as to sustain themselves. Their currencies were optimized for trade; they were less a way of saving or hoarding money than facilitating the exchange of goods between people. It worked so well that Europe saw its greatest period of economic growth to this day—measured in the prosperity of the common folk. Towns got so wealthy that they invested in building cathedrals to spur future pilgrimages and tourism. People worked less, ate more, and grew taller than ever before—and in some cases, ever since.

As the people grew wealthier and more independent, the aristocracy found itself relatively poorer and less powerful. These wealthy families often hadn't worked or created value for centuries, and needed to find a new way to dominate the masses. The first was to grant "chartered monopolies," giving favored nobles exclusive dominion over an industry. Where a cobbler may have once worked for himself making and selling shoes, now he would have to be an employee of His Majesty's royal shoe company. Individuals were denied the ability to create and exchange value on their own.

Then monarchs outlawed market money and forced everyone to use "coin of the realm." This currency had to be borrowed from the central treasury and paid back, with interest. With a monopoly on this financial technology, the aristocracy could make money simply by lending it out. Nation after nation adopted the new scheme, crushing local markets and restoring peasants' dependency on the wealthy for employment. Central currency became the new operating system for the economy, with corporations as the software that ran on it. It was a runaway program.

Growth became the new ethos and requirement for any chartered company, and the expansionist, extractive form of monopolist colonial capitalism that followed is still in full force today.
Chartered monopolies, like the British and Dutch East India Companies, traveled to the new world, killed or enslaved its people, and
extracted their resources. The church would usually arrive first,
establish contact with the native populations, and gather intelligence about them. Then the gunships, companies, and conquistadors took charge. These conquests depended on three main tenets
of corporate colonialism, which became central to The Mindset
we're contending with today.

The first was to see oneself as separate from nature, and thus
regard natives as less than human. Enlightenment thinkers, in
spite of their avid belief in the right to happiness and liberty, nonetheless based their philosophies on the empirical scientism of Francis Bacon and his contemporaries. In their schema, the New World
was "virgin land," awaiting settlement by white Europeans. "In
the beginning, all the world was American," explained Enlightenment philosopher John Locke, describing the pre-civilized "state of
nature." The Native American was to be understood as part of that
landscape, no different than "one of those wild savage beasts with
whom men can have no society or security . . . and . . . therefore
may be destroyed as a lion or tiger." While a master might have a
contract that gave him liberty to control his indentured Christian
servants, he enjoyed "absolute dominion" over his slaves, for they
were not fully human.

These vampiric practices may have undermined the regenerative properties of natural systems, but they were perfect expressions of the second main tenet of domination: extraction. For
example, when islanders began making and selling rope to the

Dutch West India Company, the company immediately sought new laws from Holland regulating the rope industry. They won a monopoly, after which it became illegal for anyone but the company to manufacture rope. The American Revolution was itself motivated by similarly extractive policies by the British West India Company. Colonists were permitted to grow and pick cotton, but they were supposed to sell their bales directly to the company at set prices. The company shipped the cotton all the way back to England where it was made into fabric or clothing, then shipped and sold it back to the colonists at a profit.

The more value is extracted, the fewer opportunities people have to create and exchange value through any means other than participating in the systems of domination and control that have robbed them in the first place. Locals can resist by adopting the violent tactics of their oppressors, but this risks infecting them with the same sensibilities. The cycle of oppression and extraction feeds itself.

Which brings us to the third main tenet of domination: the relentless pursuit of growth. Remember, this whole drive toward colonial expansion was instigated by the underlying math of interest-bearing currency. Everything is predicated on paying back more than is borrowed. This is what led us to mistake growth for economic health. The GDP, or gross domestic product, remains the primary barometer of national prosperity, even though it has nothing to do with how well people and businesses are actually doing. It simply measures the total market value of goods produced. A toxic spill is good for the GDP because we spend a lot to clean it up. Fixing a bridge doesn't increase the GDP as much as tearing it down and building a new one. Regenerating the supply of water in an aquifer doesn't increase the GDP as much as letting water become more scarce, and then charging everyone more. Besides, the financiers making loans only benefit from big new projects requiring massive amounts of capital.

The blind pursuit of growth has supported the relentless forward march of the dominator culture to this day. Corporations understand themselves as the colonizers, and the local populations into which they expand as indigenous natives to be exploited. In a landscape ruled by companies like this, it's hard to do business any other way. Entrepreneurship today has less to do with innovating a product than innovating on the business model for growth. Never is the growth itself questioned.

As a result, technological innovation has become understood less as a way to bring better, more fulfilling products and experiences to people, than as another means of doubling down on domination, extraction, and growth. The assembly line, for just one example, had little to do with making products better or even faster. Its real purpose was to minimize a company's reliance on skilled workers who could ask for fair wages. Manufacturing technology was primarily about disconnecting human labor from the value being created.

Technology companies today have inherited these same basic principles. The fact that so many of their founders are plucked from college before they've had a chance to study the history of economics, the moral philosophy of Adam Smith and John Stuart Mill, or the basics of Marxism, renders them all the more vulnerable to the dehumanizing, extractive, and growth-centric priorities of the business landscape.

They seek monopolies because that's the default structure for controlling a new market. They may use innovative technology to accomplish this, but they never challenge the underlying operating system or its demand for extraction and growth.

They justify all the resulting social and economic devastation as what economist Joseph Schumpeter called "creative destruction." While Schumpeter was quite specifically building on a Marxist idea that changes in industry can create a churn between old wealth and new wealth, the startup economy doesn't really follow this path. A handful of entrepreneurs and developers may get very rich off their ideas, but for the most part it's the same institutional investors and family funds profiting now off Google or Facebook who once profited off Intel and IBM or, before that, GE and AT&T. The media stories about the turnover of tech moguls hides the fact that the same invisible population of rich people behind them are simply getting richer.

The apps and platforms are indeed designed to disrupt markets, but primarily for the purpose of extracting wealth from the poor and delivering it to the rich. Amazon makes book publishing more "lean" by leveraging its monopoly to seize a bigger share of profits than regular bookstores did. Uber drivers make less than cab drivers and restaurants lose their margins to Grubhub. This is not creative destruction, but *destructive* destruction—all justified as the inevitable tide of forward progress.

Creative destructionists like to argue that those made destitute or unemployable by new technology simply need to learn new skills. But educating ourselves around the new needs of corporations is a dangerous game, particularly when those companies treat their employees in the fashion of early industrial factories putting people on the assembly line. Learning to code sounds like the next great American employment opportunity until companies begin outsourcing their software development to India, eastern Europe, or artificial intelligence.

The Mindset considers human beings so unnecessary, even burdensome, that the business plans for many startups are rejected if they can't demonstrate that their operations will one day be fully automated. A few employees are fine while a company is getting going, but eventually all of those skills need to be automated in order for the company to "scale" infinitely. This is why Facebook wants AIs or—at worst—its users to monitor and tag offensive posts instead of paid human employees. Any solution that involves valuing human labor risks slowing things down.

Stephen Pinker, an optimistic cognitive scientist who believes the human mind is computational, has become The Mindset's best spokesmodel for the inevitable triumph of these technocratic, market-driven solutions. The key is to keep moving forward. As he explains, "some kinds of social change really do seem to be carried along by an inexorable tectonic force." In his 2018 book *Enlightenment Now*, Pinker credits the European Enlightenment (the same one that brought us John Locke and the justification of slavery) with an aggregate decline in violence and increase in health, longevity, education level, and universal human rights.

It's a highly problematic account. First off, as David Graeber and David Wendgrow demonstrated in their myth-smashing book, *The Dawn of Everything*, the oversimplified unidirectional narrative of civilizational progress from agriculture to cities and through technology and the Enlightenment to modern society is just wrong. There have been all sorts of different city-states throughout history, with and without what we think of as technology. Even some hunter-gatherer societies had tremendous, city-size settlements with massive architectural constructions and democratic citizens' councils.

The other problem with Pinker's oft-quoted statistics on progress is that—like purist Enlightenment philosophy—they leave out what's happening in the real world. Right now, on average, human beings may live longer than they did before—but on a planet with a corresponding 58 percent decline in vertebrates and 81 percent reduction of animals in freshwater systems. He offers that "racist violence against African Americans plummeted in the twentieth century," but fails to include that incarceration rates have sky-rocketed. He writes optimistically about our ability to find new sources of water and energy, such as digging deeper under aquifers or fracking for gas. But these are really just loans against the future—a bit like citing the strength of an athlete who has been using steroids to boost their performance.

In Pinker's view, we should be happy that growth-based capitalism provides the forward momentum required to get us out of trouble. "As we have seen," he explains, "a market economy is the best poverty-reduction program we know of for an entire country." More important, according to Pinker and other techno-accelerationists, it is already too late to choose otherwise. We can't go back to being hunter-gatherers. The answer to topsoil devastated by Monsanto's chemical pesticides is new research into genetically modified food by, uh, Monsanto. There is no way out but through.

Instead of shunning this innovation, they argue, we should incentivize it through the free market. Intellectual dark web heroes such as psychologist Jordan Peterson help make the case. "What we want are just hierarchies of competence," he demands of his packed lecture audiences. "If you have a great educator or leader or thinker, you want to reward them. It's not a reward for their intrinsic being. It's a calculated move on your part to suck everything out of them that's valuable, as fast as you can." He's restating the classic Enlightenment values of extraction, hierarchy, and accelerated growth. He talks as if saving lives or the planet itself needs to be extrinsically motivated by the market, when we already know from education psychologists and scores of studies that this doesn't work. Extrinsic rewards such as cash bonuses have been shown to *demotivate* workers over the long term; conversely, a sense of connection to the work, a greater sense of meaning, or intrinsic rewards such as increased responsibility lead to better outcomes.

The Mindset holds that if there's enough profit to be made, someone will figure things out. We should stop seeing ourselves as "vile despoilers of a pristine planet," Pinker explains, and instead accept the "Enlightenment view" that all problems, if studied long enough, can be understood and solved. Environmental problems included. He's correct, at least in the sense that only an abstracted Enlightenment thinker would consider the environment a "problem to be solved" rather than a system in which we are all already enmeshed.

Proponents of The Mindset update Enlightenment notions of colonialism, conquest, and growth with Silicon Valley ones of progress, ubiquity, and scale. We're on the same, inexorable journey west into the future, only built with technology and fueled by capitalism. Any doubts or second-guessing will sabotage our "sprint" to the next "milestone" or, worse, dampen market enthusiasm for our inevitable triumph. Like one of Aristotle's heroes, we must single-mindedly follow the arc of the story to the climax.

The Mindset's version of capitalism doesn't stop at necessary growth and progress. It's about more than even just growth for growth's sake. It's reaching toward something beyond victory itself: total domination. The tech billionaires have already accumulated more wealth than they or their grandchildren could ever spend. Jeff Bezos has a yacht with a helipad that serves as a *companion* yacht to his main yacht, which has large sails that would get in the way of his helicopter during takeoff and landing. There is no such thing as enough.

This drive toward wealth and power is like a poker game where everyone stays at the table until a single player has won all the money. It's a drive toward inequality as the ultimate goal—what economists would call a Gini coefficient of 1—where just one person has accumulated everything. All the financial, technological, and cultural feedback loops in which they are participating support this singular drive. As game theorist John Nash (the subject of the movie *A Beautiful Mind*) demonstrated in his early work, the wealthier party in a transaction always has an advantage if no rules or limits are put in place to counter this effect. A game of "no limits" poker always favors the wealthier player, because they can repeatedly force their opponent to stake the entirety of their holdings. So the very existence of inequality in an unregulated market favors those with the most wealth. That's why they use that wealth to push for deregulation, which in turn wins them more wealth.

Players at this level are pursuing a very particular kind of wealth. It's not based on dividends, regenerative markets, or the circulation of money through the system. It's simple conquest and extraction. Find either a new territory to conquer and dominate, or a new technology through which to extract more than you already do. Then sell the whole enterprise before it peaks or is itself disrupted by the next new thing. And it's not all fruitless: as Steve Case showed us, the arc of crisis, climax, and exit is what got America online.

But this only works if the founders don't look back. They must rush ahead and leave everything—including us—behind. Their complete disregard for history may be the tech triumphalists' most tragic flaw. As Pope Francis explains in his trenchant critique of technocapitalism, *Fratelli Tutti*, "there is a growing loss of the sense of history, which leads to even further breakup. A kind of 'deconstructionism,' whereby human freedom claims to create everything starting from zero, is making headway in today's culture. The one thing it leaves in its wake is the drive to limitless consumption and expressions of empty individualism. . . . Some parts of our human family, it appears, can be readily sacrificed for the sake of others considered worthy of a carefree existence."

This misconception that there was nothing out there to begin with grants "developers" the freedom to destroy existing cultures, economies, ecosystems, and neighborhoods. Uber, Airbnb, and even Google see low-income residents and their neighborhoods the way John Locke saw the landscape and natives of America: as undeveloped, virgin territories for exploitation. It's no wonder their highly paid young developers use the same language when describing their search for apartments as "pioneering" new neighborhoods on the outskirts of what is normally considered "safe" for white professionals.

By combining a distorted interpretation of Nietzsche with a pretty accurate one of Ayn Rand, they end up with a belief that while "God is dead," the *übermensch* of the future can use pure reason— objectivism—to rise above traditional religious values and remake the world "in his own interests." It was actually Nietzsche's sister, a great admirer of Mussolini's fascism, who assembled the infamous text *The Will to Power* from her brother's abandoned scrawlings after his death. But his language, particularly out of context, provides tech *übermensch* wannabes with justification for assuming superhuman authority. Just as Nietzsche-inspired Uber uses its money and influence to change zoning and employment laws in its favor, Peter Thiel hears in Nietzsche a call to take the future into his own hands—in Thiel's words, "I no longer believe that freedom and democracy are compatible." This distorted image of the *übermensch* as a godlike creator, pushing confidently toward his clear vision of how things should be, persists as an essential component of The Mindset. You don't get hockey stick stock charts without such totalized, dominion thinking.

For Thiel, this means being what he calls a "definite optimist." Most entrepreneurs, he writes in his book *Zero to One*, are too process-oriented, making incremental decisions based on how the market responds. They should instead be like Steve Jobs or Elon Musk, pressing on with their singular vision no matter what. The definite optimist doesn't take feedback into account, but plows forward with his new design for a better world. It happens *ex nihilo*— literally "from zero to one."

Google's founders, Larry Page and Sergey Brin, also eschew incremental thinking in favor of breakthrough, moonshot, utterly unprecedented, game-changing innovation. With a monopolist's perspective on business, Page explained to *Wired* that "it's hard to find actual examples of really amazing things that happened solely due to competition. . . . That's why most companies decay slowly over time   Incremental improvement is guaranteed to be obsolete over time." It's not that Page is afraid of competition. Rather, if he's competing with someone at all, it means he's not in virgin territory.

This may make some sense in the landscape of unicorns and 1000x returns. How else to win the faith of venture capitalists and the necessary funds for domination if you don't talk a good, confident game? But it's a mentality derivative of the most dehumanized, disconnected, and triumphalist aspects of empirical science and Enlightenment thinking. The founder is God. He creates a new world, brings his followers through a great exodus—or exit strategy—to salvation, while the rest of humanity is left behind.

For all the claim to originality, many of these tech titans model themselves after historical figures they heard about before dropping out of school. Mark Zuckerberg is famously obsessed with Roman emperor Augustus Caesar, who is often credited with developing the network of roads and the courier system through which Rome administered its imperial expansion over several centuries. "Basically, through a really harsh approach, he established two hundred years of world peace," Zuckerberg told *The New Yorker*. (This is only true if you count "peace" as freedom from wars that threaten your own sovereignty, and exclude your empire's violent conquests of other states and peoples.) But Zuckerberg's admiration of the emperor borders on obsession. He models his haircut on Augustus; his wife joked that three people went on their honeymoon to Rome—Mark, Augustus, and herself; he named his second daughter August; and he used to end Facebook meetings by proclaiming "Domination!"

We're all certainly better off with the sole decision-maker of a three-billion-member social network modeling himself after Augustus Caesar than, say, his eventual successor, Caligula. But Zuckerberg's acceptance of Augustus's "really harsh approach" as the price we pay for the eventual stability of his empire may be too expensive. Zuckerberg's famous exhortation to his company to "move fast and break things" has won him monopolies, but has also had a disastrous impact on internet innovation, the social landscape, mental health, and the viability of democracy itself. That's what happens when you are a definite optimist, driving toward a single goal, with billions of dollars and petabytes of RAM at your disposal. Sure, Zuckerberg can later promise to give back 99 percent of his winnings to charity, but that simply proves he took too much to begin with. Imagine if Facebook had been 99 percent less destructive in the first place.

All that said, it's too easy to credit these entrepreneurs' success to their ruthless and imperious determination. Despite their patronizing language at congressional hearings challenging their tactics and power, the tech titans may be overestimating the value of The Mindset on their company's good fortune. In reality, they've just been surfing the wave of Moore's Law, the exponential growth of processing power—a technological trend well beyond their control—and mistaking their success as their own manifest destiny. But Moore's Law could be slowing down, auguring an end to automatic exponential growth. In 2010, Robert Colwell, director of microsystems at DARPA, stunned attendees of a Palo Alto tech conference when he announced, "I don't expect to see another 3,500x increase in electronics—maybe 50x in the next 30 years . . . we will make a bunch of incremental tweaks, but you can't fix the loss of an exponential."

Once addicted to exponential growth, it's hard for an entrepreneur, a company, or an entire economy to slow down. Everyone's "cap tables" are modeled on continuing expansion at ever-increasing rates. That's how the debt-based economy works. The 2008 Lehman Brothers collapse showed us what happens when the Ponzi scheme breaks down. But not to worry. While their ability to increase the raw speed of their machines may have reached certain limits, the digital environment offers them another way to go from *creatio ex nihilo* to *deux ex machina*. They just go "meta."

Exponential

WHEN YOU CAN GO NO FURTHER, GO META

I got mugged in front of my apartment building while I was taking out the garbage. I posted a note about it to the Park Slope Parents list—an online community dedicated to the health and wellbeing of local families. The very first responses were from people who were *angry* that I had posted the location of the intersection near which the crime occurred. Didn't I realize that this publicity could adversely affect all of our property values?

No, they weren't looking to sell. But the five-year "interest only" period of their mortgages was about to expire. In order to refinance at better rates and with greater equity, they needed home values to go up. They feared that a disruption as minor as "writer mugged in Brooklyn" could crash the real estate scheme on which their finances depended. That's why they cared more about the abstract market value of their homes than the quality of real life in their neighborhood.

The brittle, highly speculative, growth-dependent shell game that people were using to live in otherwise unaffordable Park Slope brownstones was itself the basis for a much bigger series of bets and abstractions on those bets. All these specious loans were part of the "subprime" mortgage market, which was itself serving as the collateral for a series of other loans, and so on, and so on. All those loans and loans on loans were packaged into "baskets" that were sold to investors. Shares in these baskets, in turn, could be speculated on through other derivatives, which could themselves be bet on—or against—with credit default swaps.

It would have worked forever, as long as housing prices kept increasing at ever faster rates, supporting all the financial instruments that were depending on them. But the inevitable crash came just a few months after my mugging. Experts differ on exactly what triggered it. Either the rise in house prices began decelerating, interest rates went up, or both. In any case, it became harder for people to refinance their properties at higher valuations. Instead, when the interest-only period of their existing loans expired, homeowners began defaulting. Then the whole house of cards came down, bankrupting nearly everyone except financial firms like Goldman Sachs, who had been betting against the very mortgage products they were selling to investors.

What was really going on here was that the real world of houses and property was unable to support so much financialization. In their race to grow exponentially forever, the markets had "gone meta" one time too many. No matter how many digital signifiers we employ to represent its value, the real world just doesn't scale forever.

A similar process of "going meta" fueled European colonialism. Under the supervision of kings and their navigators, land came to be represented by maps. Maps could be labeled, transforming land regions into territories. Places became properties. Pastures became parcels. Land was now an asset that could be bought, sold, and chartered. Back home in Europe, too, land went from feudal to market control. The aristocracy used their lands as a form of capital to trade, while the new class of wealthy merchants purchased land to earn social distinction.

Most importantly, land had gone from a living ecosystem on Earth to a more abstracted unit of exchange. The deed can be thought of as "once removed" from the land to which it refers. It's *meta*, and once things go meta, they tend to keep on going that way. Land becomes property, property becomes mortgages, mortgages become derivatives, and so on. Abstracted properties can be bought and sold by people across an ocean—like those empty mega-luxury apartments in Manhattan owned by speculators and sovereign wealth funds. They may as well be stock certificates. At each level of abstraction, a new population of owners, bankers, or speculators arrives to stake its ever more leveraged claim on whatever core asset exists on the ground.

Going meta is the American way, and a foundational premise of The Mindset. The landlord goes meta on the renter, while the bank goes meta on the mortgage-holding landlord. Each layer of abstraction—in this case, financialization—allows for growth that couldn't have been achieved otherwise. In an economy defined primarily by interest-bearing central currency, growth isn't just good but required. When growth on a particular level has reached its limits, going meta lets a lucky few scale up to the next level of abstraction.

This financial pyramid is based on an ethos of individual autonomy, even for the patsy at the bottom. U.S. presidents since FDR have promoted policies and propaganda around home ownership as the basis for achieving the American dream. It wasn't until this period that the word "home" came to mean the dwelling that one owned, rather than the town or neighborhood one came from. This was the result of a conscious effort at social programming. Worried about an influx of traumatized and potentially unruly World War II veterans, FDR hoped that home ownership and mortgage obligations would help keep them under control. As William Levitt, developer of the first planned suburb, Levittown, explained it to him, "No man who owns his own house and lot can be a Communist. He has too much to do." The Federal Housing Authority facilitated better mortgage rates for single-family homes than for shared residences, for new properties over the renovation of existing homes, and, of course, for segregated "red-lined" communities over those without clear race discrimination.

This base of homeowners served as the ground level for American consumerism, which in turn provided the engine for each new level of financial abstraction. Industrial giants from General Mills and General Foods to General Motors reaped huge profits supplying homeowners with all they needed to fill their bellies and homes. The shareholders one level of abstraction above these companies did even better. And the investors buying derivatives on those shares did better still.

By the 1980s, General Electric CEO Jack Welch recognized the pattern, and its implication: get as far up into financial abstraction as possible. Like any company selling big-ticket items, GE had a capital services division to help finance purchases—originally intended as a way of reducing pain points for customers. Yet Welch observed that he was making less money selling washing machines to people than he was lending them the money to purchase the washing machines. When he was manufacturing washing machines, his profits were limited by real-world frictions such as the cost of materials, labor, and shipping. When he was selling loans, he could make money like magic. It was all just numbers, which could scale frictionlessly. So Welch went on a mission to sell off GE's productive assets and pivot entirely toward finance. *Harvard Business Review* extolled the virtues of Welch's new strategy, business schools taught it, and other corporations imitated it. In their effort to become more like banks, businesses cannibalized themselves, liquidating any divisions that were on the ground actually creating value.

Neither GE nor any of these companies had any true expertise in the financial services industry, however. So, when the easy money ended in the financial crisis of 2007, they were left in a much more precarious position than real banks. Jack Welch quickly realized there was no turning back, and jumped ship. After laying off tens of thousands of manufacturing and engineering employees, he retired from GE in a golden parachute, and his successors were left to rebuild the vanquished industrial, consumer, and aerospace divisions. GE eventually sold off most of its financial services, finally splitting off its credit card company, Synchrony, in 2014.

But General Electric's more fundamental challenge—the one Welch never adequately addressed—was that the real world of houses, airplanes, and industrial activity couldn't scale in the ways capital required and investors were demanding. At some point, manufacturing hits the hard limits of human labor and physical matter itself.

The digital realm appeared to solve this Industrial Age problem by transcending the laws of physics. In his 1995 book *On Being Digital*, Media Lab founder Nicholas Negroponte announced to the world—and to businesspeople in particular—that "bits" had come

to rescue us from the tyranny of atoms. Industrialism was limited because "world trade has traditionally consisted of exchanging
atoms." Now that we were moving into a digital age, however,
the limits of the physical world no longer applied. "A bit has no
color, size, or weight," he explained, "and it can travel at the speed
of light."

Technically speaking, none of this is true. For a bit to be a bit
it has to be recorded somewhere—on a disk, a piece of paper, a
RAM drive, a synapse . . . It does exist in the real world, and is
constrained by the limits of physical reality. Ask anyone who does
high-frequency trading stock trading, and they'll tell you that the
distance of the server from those at the trading desks makes all the
difference in whose electrons arrive first, and which firm gets its
orders executed at the best prices.

Still, the compelling idea here is that bits are relatively abun-
dant compared to matter. Just as the word "cat" is infinitely more
replicable than a real cat, digital representations are just symbol
systems—abstracted 1's and 0's—that are identical copies of each
other, and endlessly replicable. This scalability of data renewed
aspirations for an infinitely expanding market. The digital revolution would take place on an abstracted plane above and beyond
the constrained world of people, places, and things, allowing businesses to grow simply by going meta. This is when *Wired* ran its
"Long Boom" cover story, arguing that thanks to the potential for
infinite scalability, the global economy would now grow exponentially, forever. Even Fed chair Alan Greenspan signed on, admit-
ting that the normal rules of economics no longer applied and that
we were in a "new paradigm"—a dimensional leap in the behavior
of capital itself. Money was another form of data, and data another
form of money.

AOL and the other dotcom casualties appeared to disprove the new thesis. But these online shops and services were really just version 1.0 of the digital economy. They were all dependent on human customers with attention spans, and physical equipment that cost money. AOL wasn't really a digital company; it was a dial-up service, requiring a separate modem and phone line for every user who wanted to log on. Dotcom companies may have been accessed through websites, but they still sold products that had to be delivered on real planes and trucks. Just because a business conducted its business on the internet didn't make it truly digital.

The companies that survived the dotcom boom had one thing in common: they had gone meta. Tim O'Reilly, the technology book publisher, called it Web 2.0. He said that Web 2.0 companies, like Google and eBay, treated the web as a platform, and leveraged the activity of users rather than spending money on their own people and merchandise. Unlike Yahoo, Google didn't hire human employees to create taxonomies of the web; it used algorithms to catalog everyone's existing hyperlinks and then organized them into a searchable database. Unlike the many storefronts that opened online, eBay developed an automated platform that connected sellers and buyers. Web 2.0 companies and projects, from Wikipedia and Blogger to Sourceforge and iTunes, relied on peer production. They were themselves meta operations that simply aggregated everyone creating value at the level below.

What made a business truly digital was whether it could rise one level above the competition. Each new level amounted to an exponential leap, from x to x-squared to x-cubed and onward. A travel platform (Expedia, Travelocity) goes meta on the airlines, aggregating the data from all their websites to show you the best prices it can find. One level above that, an aggregator of those aggregators (Kayak, Orbitz) can show you which aggregator is doing this the best. Don't focus on the content, experts like

O'Reilly insisted, but the platform on which everyone posts the content. And if there are already a bunch of platforms, become the platform of platforms. "The medium is the message" became the business mantra for The Mindset, while Marshall McLuhan himself earned a posthumous place on the *Wired* masthead as the magazine's "patron saint."

According to Peter Thiel, any new business idea should be 10x better than what's already out there—literally, an order of magnitude better. Borrowing from his former Stanford philosophy teacher, René Girard, Thiel believes that "competition is for losers." Everyone in the world is engaged in a simple game of copying one another, or what Girard calls "mimesis." While this is a great way for kids to learn from their parents, among adults it creates a culture of competition, where everyone covets what their neighbors have. They go on like this until the competition gets so extreme or even violent that they eventually choose a scapegoat (Jews, immigrants, queers, or even an individual) on whom to blame their conflict. Violence then relieves the tension, and the competition starts over again. (Girard and Thiel believe that Christ was meant to end this cycle of violence by serving as the last and ultimate scapegoat. The crucifixion and resurrection of the son of God could liberate humanity from the cycle of violence—if only people could be encouraged to believe in the myth as literal truth.)

The implication for businesses, though, is to avoid competing with everyone else and instead innovate on the next level. We are to achieve this by maintaining a "fidelity to an event"—a singular devotion to a future that others can't yet see. Thiel saw this most clearly in Mark Zuckerberg's Facebook: instead of competing to build the best website or personal homepage, Zuckerberg leveled up to build the platform where people and companies can do this. Rather than imitating, he transcended the game. He took that exponential leap, one order of magnitude above mere mortals and into the realm of success, autonomy, self-determination, and salvation. Amazingly, now that Facebook's business model has come under scrutiny, Zuckerberg is at it again, going meta on the net by rebranding his own company as Meta. He is preemptively attempting to aggregate yet-to-be-invented virtual and augmented reality technologies into a single "metaverse" over which he will preside from one level up.

The postmodern style of business warfare, where companies each seek to leapfrog one another's paradigms, takes place in the financial markets capitalizing them, as well. Investors race to invent new derivatives and meta-derivatives capable of subsuming or aggregating those that came before.

But the real leap came once traders replaced themselves with algorithms capable of aggregating data from all the trading platforms and executing high-frequency trades at a rate and volume beyond the cognitive capacity of hundreds of human beings. These derivatives markets quickly outpaced the traditional trading activity on the stock market. Derivatives trading became so dominant that the New York Stock Exchange was actually purchased by its derivatives exchange in 2013. The stock market—already an abstraction of the real marketplace—was swallowed by its own abstraction. Meanwhile, still more technologists attempt to level up again and again by selling the trading algorithms, the machine learning to devise those algorithms, or the platforms to support machine learning. Each level of abstraction begets the next.

Yet they all depend on the initial contention of the digital revolution that anything that matters can be digitized. Just as maps abstracted land into monetizable parcels, computers convert things into their digital counterparts, rendering them grist for the exponential mill and supporting the underlying need under capitalism for money to grow. Nowhere has this been made more clear than in digital's replacement for central currency—crypto.

Initially conceived alongside Occupy Wall Street, the bitcoin protocol offered a way for people to authenticate transactions without involving banks, fees, and usurious intermediaries. But, just like the monarchs behind central currency, speculators were less concerned with facilitating transactions than profiting off them and raising the price of the Bitcoin token. Millions of computers around the world now have no other purpose than to prove the value of Bitcoin by spinning their cycles and spending electricity on purposeless calculations—amounting to a bit more than the total energy consumption of all of Sweden. We are quite literally burning the real world to prove the value of digital symbols— feeding reality to its more scalable digital counterpart.

For holders of The Mindset, all this wasted power is like the first stage of the rocket ship taking them to the next level. It spends a lot of fuel before it is discarded and allowed to crash back to the planet while the astronauts continue on their journey. Don't look back, just look forward. Of course, the real money will be made by the companies who go meta on this trend. While crypto investors gamble through investing or eke out small margins by mining coins themselves, smarter players look to become the casino and build the exchanges where all this trading takes place. In April 2021, Coinbase was the first of these exchanges to go public, with an IPO valued at about $100 billion. As if aware that someone had gone meta on their holdings, institutional crypto traders began cashing in their tokens that week, leading to a crash in crypto currencies.

When value is best created by going meta, the data points *about* our world tend to become more important than whatever is in the world itself. Pork belly futures are more fungible and scalable than the actual pork bellies. Data is just cleaner, lighter, and faster than its real-world analogs. May as well convert everything to digital. We are each becoming more valuable as data than we are as real-world consumers or even humans. This leads to a disconnection between benefits and incentives. The companies behind our activity trackers and exercise apps often make more money off our data—usually anonymized—than off making us healthier. Our social networks can make tremendous profit off a teenage girl's data profile, even if the platforms themselves make that girl more likely to self-harm or worse. The cloud doesn't care. The teenage girl has ceased to be a girl. She has become pure, abstracted data. It's digital heaven for those who know how to ascend, and something else entirely for those of us left behind.

In fact, the most devout holders of The Mindset seek to go meta on themselves, convert into digital form, and migrate to that realm as robots, artificial intelligences, or mind clones. Once they're there, living in the digital map rather than the physical territory, they will insulate themselves from what they don't like through simple omission. Just as our proprietary GPS maps don't show us the restaurants that refuse to advertise on the platform, the digital landscape to which they have migrated will be free of poverty, pollution, and whatever else the rest of us have to deal with.

As always, the narrative ends in some form of escape for those rich, smart, or singularly determined enough to take the leap. Mere mortals need not apply. I got into a heated discussion about this with transhumanist Ray Kurzweil. Being interviewed for a TV show, Ray and I had each just shared our most optimistic visions for the ways technology would redefine what it means to be human.

For me, it was about enhanced connectivity and maybe a newfound appreciation for the non-technological, sacred weirdness of corporeal existence. For him, it was about transcending mere mortality and merging with the machines as pure data. He explained that within just a couple of decades (and he's now been saying this for a couple of decades) human beings will achieve immortality by uploading their minds to the cloud and downloading them into fresh hardware. Everything about us that can be converted into data will be preserved. Anything that cannot, well, that stuff isn't real, anyway.

I made an impassioned plea for aspects of the human experience that cannot be transferred to the cloud. "What about the soft, squishy stuff?" I offered. Human beings can embrace and sustain paradox over time. Not everything about us can resolve to a one or a zero.

Kurzweil called that "noise." He explained that my perspective was far too human-centric. Information was really in charge here, evolving ever since the formation of the universe toward higher states of complexity. Once computers can support greater complexity than the human brain, information will inevitably migrate from our biological processors to the superior digital ones, now presumably being engineered by his team Google, where he currently serves as a senior technologist. After that, human beings will only be important insofar as we are needed to service the machines. Then, we must learn to accept our obsolescence. If we want to be a part of the future in any form, we need to drive with singular vision toward "the singularity" itself, and offer up whatever about us can be converted to pure data.

Kurzweil's vision is a platform-agnostic understanding of life, mind, and information. The data we contain—the software we run—is equally at home on a silicon chip as it is on the wetware of brains. As Google co-founder Larry Page puts it, human DNA is just "600 megabytes compressed, so it's smaller than any modern operating system . . . So your program algorithms probably aren't that complicated." This model of human biology is as reductive as Dawkins's contention that "life is just bytes and bytes and bytes of digital information." Just as Francis Bacon and the early empirical scientists denied any aspect of nature that could not be quantified, today's digital reductionists would have us deny any aspect of the human experience that cannot be quantized as code. Everything can be represented as symbols. It's all just information. Nothing weird, wet, or truly wild. The ultimate nerd religion.

By refusing to recognize anything that can't be quantized to a one or zero, this analysis misses everything in between. It depicts an autotuned reality, where every note must be averaged up or down to the nearest quantized notch. The subtleties of a vocalist's interpretation—what true music lovers listen for most—are discounted as "noise." The emphasis on life as a form of code also ignores the context and culture in which that life is unfolding. More nuanced scientists recognize that DNA is important, but it's not even half the story of how a life form expresses itself. Rather, DNA is a set of potentials entirely dependent on the protein soup in which it finds itself. Our bodies and minds may be no more the tool for the preservation of DNA than DNA is a mere scaffold for human and other life to express itself.

The reduction of reality to information and humans to genotypes all-too-conveniently dovetails with capitalism's imperative to render everything into a suitable form for the marketplace. Everything is data, and everything has a price, and everything can scale. The described, codified object is all that matters; anything else falls away like junk DNA, inferior species, or the majority of human beings. The wealthy technologist makes it into the cloud, while the masses are left behind competing against one another in the realm of matter. Like Christ or any other saved figure, only the fully encoded individual can be transubstantiated to the next level.

So goes the atheistic eschatology of The Mindset.

8

Persuasive Tech

IF YOU COULD JUST PUSH A BUTTON

It's January 6, 2021. I'm on a Zoom call to talk with some tech developers about a new social network they're building. Well, it's not exactly a social network, at least not according to them. It's something better, different, and less manipulative, that uses the blockchain to reward people for the content they create, the attention they pay to other people's content, and the attention their attention or content creates for other people's content. (Plus, a piece of future attention that those recommendations initiated.) They don't call it "content," either, but use a word from Sanskrit or Zen that means, well, something like content.

They seem like nice enough guys (I know, I keep saying that). One is just graduating from Stanford, and the other two are going to leave their jobs at Twitter and Facebook to start this new, better, healthier, decentralized platform, once they launch their token and have enough money to hire themselves for real.

"Shit, check this out," the guy from Facebook suddenly says, rescuing me from having to weigh in on their business plan. Apparently he's been multitasking. He shares his screen with us: video footage live-streamed from four different locations within protests at the Capitol. The one on the top left shows the crowd breaking through the barricade and swinging poles at the police. "Fuck."

"Fuck," the Twitter guy agrees.

It's an unsettling, emotional moment, yet our conversation got strangely abstracted—as if The Mindset's defense mechanism against trauma had kicked in. The Stanford student, the most mission-focused of the trio, tied it back to the value proposition of their platform.

"This is why we need systems that amplify signal over noise," he explained. "These people are all victims of fake news and networks optimized for sensationalism. Imagine if we tuned the network to promote cooperation and consensus."

"How did this guy know to assemble all these feeds?" the Facebook guy asked about the page we were watching, which had aggregated six different live-streams. "He must have been tipped off that something was coming."

"Of course they did. This was all planned," added the Twitter guy. "This is the beginning of a civil war."

"Or maybe the end of one," I added. Honestly, I'm not sure what I meant by that. I suppose I was suggesting that this was the same Civil War that America's been fighting since Lincoln's day, having never resolved fundamental questions of race, sovereignty, and entitlement.

"They're crazy," the student said, no doubt musing on the off-campus reality he was about to re-enter. "QAnon conspiracy people are going to cost us democracy."

We watched in silence as one of the video streams followed along with a group of protesters, through the police line and into the building.

"What if you could just make all those QAnon people disappear," the Twitter guy said. "Would you do it?"

"What do you mean, 'disappear'?" asked the student. "You mean kill them?"

"No. Not like that," Twitter explained. "I just mean, well, if you could just push a button and have those people not exist anymore. Like, all the people who believe that stuff just don't exist anymore . . ."

"And all the logical paradoxes of erasing their existences from the timeline are automatically resolved, as well?" added Facebook.

"Yeah," Twitter said. "If you could just push a button and have them not exist, would you do it? For the sake of democracy?"

"Or better," offered Facebook, "you push a button and they no longer believe the crazy stuff. Everything else about them can stay the same. They just stop believing the crazy stuff."

"For the sake of democracy," I added, wryly.

"The world would be a better place," the Stanford student mused.

I guess you can take the engineer out of Facebook, but you can't get Facebook out of the engineer.

Don't worry. Such technologies do not exist. This was a thought experiment, in the heat of a scary moment. But it's emblematic of the way those trapped in The Mindset seek to change people—to make them more compatible with a free, open, happy, progressive, and tech-enhanced society, and to do so at a great remove. You don't have to get in anyone's face, confront them directly, or even hear what they're really saying. Just push a button and make it go away. Swipe left.

This impulse to manufacture consent and exercise social control from above has informed media and technology practices for a very long time. And, perhaps surprisingly, it was born not in the conference rooms of ruthless Madison Avenue advertising firms, where it was later practiced, but out of the musings of one of Woodrow Wilson's most progressive advisors: political commentator, co-founder of the *New Republic*, and the father of "public relations," Walter Lippmann.

Once a member of the New York Socialist Party, Lippmann was concerned that people are more apt to believe and react to "the pictures in their heads" than whatever is really happening in the world outside. Journalism and other media inserts what Lippmann called "a pseudo-environment" between us and the environment in which we are actually living. This pseudo-environment, in turn, stimulates us to act in ways that really *do* change the world. Like the people in the allegory of Plato's cave, we're responding to what amounts to shadows on the wall, and this makes us particularly vulnerable to dangerous despots and demagogues who can create the most compelling pictures.

Where the people reforming today's social media platforms may hope to mitigate the influence of alt-right extremists and conspiracy theorists on our beliefs and behavior, Lippmann and Wilson were concerned about early-twentieth-century nationalists who sought to keep America isolated and turned inward. Their concerns felt justified. They had lived through Teddy Roosevelt's failed effort to engender a progressive populism. In a fashion vaguely similar to Trump, Roosevelt rose to power by articulating the complaints of the working poor against corporate elites. He demanded the press expose corporate corruption and the anger of the common people, but all this "muckraking" only frightened the middle class reading about angry crowds in the daily newspaper. Otherwise compassionate progressives found themselves more concerned about taming the mob than addressing whatever underlying issues were leading to the unrest.

Lippmann and his contemporaries had come to understand the public through the lens of French sociologist Gustave Le Bon's immensely influential and frightening book, *The Crowd: The Study of the Popular Mind*. He argued that the crowd subsumes individuals into a new psychological entity, capable of terrible acts. The mob was dangerous, violent, and a threat to the social order. If it truly got out of control, it could elect a demagogue, undermine democracy, scapegoat a group, or come for us all. As a result, Lippmann had no misgivings about steering the mob toward his idea of the mob's best interests.

Woodrow Wilson had run on a nationalist peace platform, but once he became president he felt he needed to get the public to support American intervention in World War I. Lippmann advised Wilson to create a pro-war propaganda committee, which became the Creel Commission. Its job would be to insert new pictures into people's minds and influence what Lippmann dubbed "public opinion."

The only way Lippman could justify such manipulation was to fall back on his foundational assumption that people living in a modern media landscape are utterly incapable of knowing what is really going on. Helplessly responding to whatever "pseudo-environment" our newspaper or radio station created for us, we were making choices based on fictional pictures drawn by self-serving people and institutions. This could lead us to vote for the wrong people, and support policies that hurt us and the nation. To make matters worse, according to Lippmann, it would simply take too long to teach people how to think for themselves. Neither our education system nor our institutions of journalism were up to that challenge.

Instead, Lippmann believed the government should install a "board of impartial experts"—scientists, statisticians, doctors, and so on—who could serve as "independent givers of fact." They would report directly to the elected officials, who could develop policies based on reality and reason. Then, these policies would be sold to voting citizens through the very best "education," or what became known as public relations. Propaganda is what the *other* guys do. According to Lippmann, a properly functioning democracy depends on a benevolent elite to determine our society's best courses of action and then use whatever media tactics are at its disposal to "manufacture consent" from the public.

Once the tactics were out of the bag, less scrupulous practitioners began deploying them not just for the government but also for corporations. For instance, when the duly elected government of Guatemala began instituting policies to protect workers and landowners from exploitation by the United Fruit Company, Lippmann protégé and former Creel Commission member Edward Bernays invented the mythology that the country's new president was soft on Communism, and possibly even a collaborator with the Soviet Union. He staged scenes for newsreels and other propaganda to garner support for a U.S. invasion. It worked, and under the guise of liberating Guatemala, the United States restored an oligarchic regime, slave labor, and private control of agriculture.

Bernays wrote the book on propaganda—literally, it was called *Propaganda*—in which he explained that the manipulators of public opinion are the true, invisible power in any society. The masses are too stupid to make decisions for themselves, anyway, so their rise to power in a democracy must be steered by propaganda, a "mechanical, advanced, and necessary" science of population control. At the time, many journalists and politicians spoke out against Bernays's tactics and beliefs. He and his colleagues, however, saw themselves as conducting an essential social service. The elite, and even progressives among them, feared the potential power of the unchecked crowd to wreak havoc. They witnessed the madness of crowds in Nazi Germany and the Stalinist Soviet Union alike, and wanted to prevent such crises from happening in America, much as Hillary Clinton feared the "basket of deplorables" who might elect a demagogue as president, or those young technologists feared the insurrectionists at the Capitol. If only one could push a button and make all that go away.

The emerging science of psychology seemed to offer such buttons. On one end of the new discipline's spectrum was Bernays's uncle Sigmund Freud, who offered insights into personality, primal emotions, and the subconscious. Psychoanalysis meant there was a "self" within oneself, to which a propagandist or marketer could speak directly through symbols and imagery. Bernays used such methods for his infamous "torch of freedom" campaign, for which he hired models to march in New York's Easter Sunday Parade while smoking. This was meant to break the social taboo against women smoking in public by associating cigarettes with liberating women's repressed oral fixation and sexual desire.

On the other end were the behavioral scientists, like Harvard's B. F. Skinner, who likewise believed that the "freedom" treasured by people in democracies was entirely illusory. Rather, like rats in a maze, human beings were merely responding to rewards and punishments doled out by those in control or by the environment itself. We are conditioned to eat berries and run from lions the same way we are conditioned to stop at red lights, genuflect at a church's altar, or order a Big Mac.

The famous Skinner Box, where an animal pushes a lever to get food, became a metaphor for all sorts of "operant conditioning" performed on human beings in casinos, shopping malls, and other spaces where environmental triggers and rewards can be totally controlled. Studying humans in these locations became commonplace, with behavioral engineers using surveillance cameras to track consumers' movement patterns, likelihood of inspecting merchandise, and response to changes in layout, color, or lighting. By subjecting people to a "technology of behavior," we could not only create better gamblers and consumers, but better and more cooperative people.

Over time, different anthropologists and social scientists extended these two basic approaches to controlling human behavior. Gregory Bateson and Margaret Mead, perhaps most importantly, applied behaviorism not just to individuals but to whole societies. As Bateson put it, we now understood that the individual was no more than "a servosystem coupled with its environment." He and Mead believed that their new theories of "systems" could be used to "engineer" a new humanity. "How would we rig the maze or puzzle-box so that the anthropomorphic rat [the human being] shall obtain a repeated and reinforced impression of his own free-will?"

Bateson and Mead believed that a world filled with screens could meet that challenge. Putting screens in stores and malls would allow social engineers to pass people off from screen to screen, amazing them with new possibilities and offering them more choices. Consumers would be free to choose from among dozens of different laundry detergents, cereals, and toilet tissues—even if they were all manufactured by the same two or three companies. What mattered was the experience of choice, and the stark contrast between America's freedom and the Soviet Union's restrictions.

Again, it didn't come off as nefarious at the time. Rather, in a precursor to New Age spirituality, Bateson hoped that a world filled with screens, in which people were surrounded by various cues and feedback, would allow us to reach a state of shared consciousness. We would liberate ourselves from obsolete notions of God, and all become part of the same "supreme cybernetic system" he called Mind.

Throughout the 1950s and 1960s, government and corporate leaders alike hoped that computers would offer new ways of measuring public opinion and then developing appropriate "mass communications" strategies for controlling all these people. Data scientists at companies from RAND to Simulmatics sought and failed to predict and steer the behavior of consumers and voters. It wasn't until the first intentionally "sticky" websites in the mid-nineties—websites designed to keep users from surfing away—that digital technology provided the sort of controlled environment and live feedback mechanisms required to do operant conditioning en masse.

Websites, video games, and smartphone apps all serve as virtual Skinner Boxes, giving developers the ability to build in operant conditioning routines to modify human behavior. As I argued in my book *Program or Be Programmed*, software companies are no longer programming computers; they are programming us people. Notifications, swipes, Likes, and "leveling up" were all developed and optimized for their ability to trigger dopamine releases on cue and foster compulsive behaviors. Developers also leveraged the Freudian section of the psychological toolkit, appealing to our tribal instincts with features like "groups" on Facebook or "guilds" in World of Warcraft. Social networks exploit our innate "fear of missing out" by intentionally showing us pictures of our exes having fun, parties we didn't go to, and job promotions of others.

This is not just coincidence, or some circumstantial byproduct of how these platforms function. This is the science of designing for behavior change, or what Stanford professor B. J. Fogg calls "captology." The Fogg Behavior Model (or FBM, as it has become known and trademarked) seeks to encourage certain behaviors by lowering obstacles, increasing motivation, and then "prompting" the user at just the right time for them to take action. Fogg's book on the FBM, *Persuasive Technology: Using Computers to Change What We Think and Do*, makes no secret of his research findings, and has become required reading in the user interface departments of most tech companies. Compliance engineers have used the FBM to develop addictive algorithms like the ones in Las Vegas slot machines, to create suggestions for new contacts on LinkedIn, to design the "infinite scroll" on Facebook, to reinforce extremist channels on Twitter, and to devise the "streak" feature on Snapchat where kids are rewarded for making contact every day. Thanks to data mining and machine learning, technologists can use computers to operate people.

Fogg is often villainized for having hacked human behavior in order to systemize the dark art of persuasive technology. I've done so myself, casting Fogg as something of a Dr. Evil at the heart of Stanford's most overtly hubristic, anti-human laboratory, arming soulless young technologists with the tools they need to drive civilization off a cliff. After a Netflix documentary called *The Social Dilemma* came out, however, in which many of the tech industries' worst offenders each blamed Fogg's courses and lab for inspiring their most manipulative tech innovations, I started to have my doubts.

I emailed with Fogg himself, researched his background, and came to see him less as a cynical Bernays employing an anything-goes strategy against lowly humans, and more as a well-meaning, if naive, Walter Lippmann trying to help people live a bit better. Many of the technologists blaming him for their turn to the dark side had never even studied with Fogg or worked in his lab. Fogg told me that his classes were always about using the behavioral model to help people achieve their own goals. It was the aspiring tech bros who kept wanting to apply the model to addiction, surveillance, and control, and who are now blaming him for giving them the tools to do so. Fogg says he repeatedly warned his students not to succumb to the temptation to use these powerful tools to manipulate people.

More recently, in the face of widespread criticism, Fogg has taken pains to stress that his work only be used for good: "the purpose of Behavior Design is to empower you to create solutions to help people with positive behavior change" and to "[help] people succeed and feel successful at doing what they already want to do." Of course, this begs the question of what constitutes "positive" change, who gets to make that evaluation, and who determines what it is that a user already wants to do. Lippman's "council of experts"? Moreover, it accepts the underlying premise that we can make people better—or, at least, make them make better choices for themselves—by programming their behavior with technology.

So, we have apps that gently "nudge" us to eat better, take breaks from work, exercise, or even text our spouses so they believe we're thinking about them. The logic is that after a period of nudging, we will be trained to do these behaviors ourselves. Researchers at the University of Zurich are developing a smartphone app to help people change not just their behaviors, but their personalities. Chatbots engage with users on a daily basis to increase positive traits like openness, conscientiousness, sociability, and considerateness.

Gamification—the application of game dynamics to work and other human activity—is being used in a wide variety of unlikely places to increase whatever metrics are being sought. Amazon incentivizes productivity with a game called MissionRacer, in which warehouse employees advance their virtual cars around a track by sorting and packing boxes properly. Many organizations are looking at using gamification to promote environmentally friendly behavior, but—as tech critic Evgeny Morozov points out—such efforts get people to engage in behaviors with no understanding of why or how they matter.

Even many of those who have dedicated themselves to lessening the negative impact of manipulative technologies on people and our society propose solutions that are entirely informed by The Mindset. On one level, it's just about using technology to mitigate the effects of technology. So, if we are addicted to our smartphones, we install an app that nudges us to look up. If we are being made anxious by social networks or 5G radiation, we affix electrodes to our skulls and recalibrate our brains with "transcranial direct current stimulation." What one technology damages, the next technology repairs.

Nir Eyal, an "applied consumer psychologist," plays both sides of the table. His first bestselling book, *Hooked: How to Build Habit-Forming Products*, adapted B. J. Fogg's FBM into an even simpler, marketing-friendly framework Eyal calls the Hook Model. It's all just habits, triggers, and variable rewards, embedded into the way an app or technology works. His main idea was to extend Fogg's techniques into a feedback loop or "hook cycle," so that the more people use a product or program, the more addicted to it they become.

Four years later, after Trump's election spurred a widespread backlash against addictive and polarizing technologies, Eyal wrote the opposite book, called *Indistractable: How to Control Your Attention and Choose Your Life*. In this one, he gives us hints on how to take back our attention from the very companies he taught how to capture it in the previous volume. In the manner of a weapons dealer, Eyal sells munitions to both sides, profiting off the escalating arms race between people and the technology companies that would control them. When I challenged him on this at a conference in New York, he told me such concerns were "destructively Marxist." Instead of protecting people from the engines of capitalism or the wiles of digital marketers, he explained, we should be getting them in the game.

Even the tech industry's less cynical reformers only emerged after Trump was elected president. It was as if they suddenly realized what can happen if these manipulative algorithms get out of human control. The Center for Humane Technology seeks to undo the negative impact of algorithms by "upscaling humans" to cope more effectively with them. Funded by billionaire Deadhead and early Facebook investor Roger McNamee, as well as a handful of other tech bros who are now feeling ashamed of themselves for the platforms they've built, it's a well-intentioned but problematic effort. Red flags abound, from the Center's association with the World Economic Forum Global AI Council to its willingness to accept social media companies' claims about their technology's power over us at face value. Further, many of these new tech reformers have yet to divest from their holdings in Google, Facebook, and others aiming at our brainstems—companies they now claim are "as big an existential threat to humanity as climate change."

The group painfully lacks a structural critique of the market economy. Most of them appeared in the Netflix documentary *The Social Dilemma*, which was widely acclaimed for its startling admissions by members of the tech industry, as well as its fictional movie-within-the-movie about a family devastated by its use of social media. Naomi Klein told me that she screened the movie to her undergraduate students at Rutgers. Watching the Center's leaders opine about the dangers of platforms, the students said, "They're willing to see everything except capitalism."

Instead, McNamee and others blame B. J. Fogg for teaching them such techniques, and share how they wouldn't let their own kids use the apps they've built. While they may be crying all the way to the bank, these millionaire turncoats do a good job of explaining how their platforms surveilled users and then leveraged the information they collected to turn people into more extreme versions of themselves.

Of course, most of them were making arguments lifted from the works of people like Sherry Turkle, Cliff Nass, Howard Rheingold, Andrew Keen, Evgeny Morozov, Astra Taylor, Richard Barbrook, Jerry Mander, Cory Doctorow, Marina Gorbis, dana boyd, Nick Carr, Mark Bauerlain, and even Raffi. Tech critics have been writing about the impact of social media manipulation on our psyche and society for decades. It's great that the developers responsible for these misdeeds are finally agreeing with these assessments, even if they need to feel as if they've discovered the downsides all by themselves—like brand-new intellectual property. That's the way of Silicon Valley.

The bigger problem with these would-be reformers ignoring their influences, however, is that they deny themselves any theory of change or social practice. They miss out on the lessons of history, including the mixed legacies of Lippman, Bernays, Bateson, and Mead. They're destined to repeat the same, well-intentioned, mistakes.

Which is what they're doing. Their orientation is all wrong. Like Walter Lippmann changing public opinion for the public's own good, or B. J. Fogg using technology to make us choose healthier behaviors, the humane technologists still want to use technology *on* people, only toward more beneficial outcomes. In their language, they mean to "upgrade humanity" before things get truly out of control. This, more than anything, is their great fear. *The Social Dilemma* shows how technology used improperly can radicalize people and send them into the streets, with a fictional story threaded through the documentary tracing the way social media algorithms seduce a young man toward extremism. At the nightmarish, unresolved climax of the show, the protagonist in the fictional thread is swept up in a violent mob looking for vengeance.

It's that mob to which the technologists are responding most. The mob at the Capitol, the mob that elected Trump, and the mob that will storm their compounds. The wealthy technologists jumping on the humane technology bandwagon today may be less concerned about the impact of their platforms on people than the potential impact of those people on their own privilege and safety—especially if they figure out what has been going on all this time. As Peter Thiel's philosophical guide René Girard would put it, the angry mob, whipped up into a mimetic frenzy, will eventually look for a scapegoat.

If only there were a button one could push to make them go away.

9

Visions from Burning Man

WE ARE AS GODS

Y ou're watching a TED Talk. It doesn't matter which one.
Really, with few exceptions, it doesn't.

There's some guy standing in the trademark circular patch of
red carpet, telling you that everything you know about the world
is wrong. He used to think that way, too, until he had an epiphany
that turned it all around. He had the ultimate counterintuitive
insight, and realized it's not *this* way at all, it's *that* way. Black is
white and white is black. Up is down and down is up. Or, if his
insight is truly unique, left is right but—right is *also* right.

Just look at his slides, listen to his story, and allow him to negate
your felt and lived experience with his new big picture for how
things are and, more important, how things could be. Gaslighting
for the greater good. Let me uplift you from material reality, just
for a moment, so you can see the world from up here in the Pla-

tonic realm of "ideas worth spreading." Spreading, and funding. This is the idea that will change everything, for everyone, all at once, and once and for all.

There's a shape to these talks, optimized for both dramatic effect and venture funding. Like a *Shark Tank* pitch for technologies that address UN Sustainable Goals, TED epitomizes The Mindset's approach toward making the world a better place.

1. Struggle with a "wicked" problem in a conventional fashion.
2. Take the red pill to see reality in a whole new way.
3. Come back to the problem with a novel engineering solution.
4. Scale that technology globally and exponentially, saving the world from its own darker nature.

In what might be The Mindset's greatest crime against the human project, these totalizing solutions perpetuate the myth that only a technocratic elite can possibly fix our problems. They distract and discourage the rest of us from making substantive changes to the way we live, and divert limited funding to moonshot boondoggles—all while making the wealthiest even wealthier. They solve for humanity, as if we humans were the problem.

Like the birth of the internet itself, the life cycle of a techno-savior tends to begin with a psychedelic initiation, where the hero takes the red pill. This often happens at Burning Man, which has, since its humble origins as an impromptu summer solstice ritual for a couple of dozen people, grown into a desert festival with an attendance of over 70,000 burners. Instead of small tents and sleeping bags, participants—particularly the wealthy ones—now arrive in air-conditioned RVs with servants and chefs. While opinions vary on whether Burning Man has stayed true to its original ethos, the festival remains highly psychedelic, and has made the taking of acid, mushrooms, or even stronger entheogens into something of a rite of passage for the would-be enlightened tech executive.

For twenty-first-century Silicon Valley operators, these psychedelic initiations serve a purpose analogous to the way alcohol was used by media and advertising executives in mid-century New York. Getting drunk and harassing women was not just a facet of chauvinist work culture, but a way for an executive to prove he had no scruples about fucking over consumers, either. Psychedelics are a way for modern-day tech executives to show they are willing to reformat their own cognitive hard drives, and daring enough to apply those insights to the world at large. They are ready to reprogram humanity.

The psychedelics and weirdness are just a means to an end. As Google's Eric Schmidt put it, "It's well documented that I go to Burning Man. The future's driven by people with an alternative worldview. You never know where you'll find ideas." This style of engagement and exploration has grown so intense that A-list executives have created a version of Burning Man just for themselves—without the throngs of artists, musicians, and commoners who are attending for the experience of collective creativity alone. The Further Future festival of 2016, for example, adapted the aesthetics of Burning Man into an expensive alternative where entrepreneurs could do the same psychedelics but under luxury conditions, and with the express purpose of making deals. Execs in attendance at the 2016 iteration included Schmidt, Facebook's Stan Chudnovsky, and Clear Channel CEO Bob Pittman.

Billed as "a shared experience that's beyond our future," Future Festival's pretense is that the wealthy psychedelic elite who make it out to Native American territory for this party are exclusively capable of solving the world's problems. As co-founder Robert Scott told the *Guardian*, "It's important what we do here. That's what we keep saying. We're shaping the Future. These are the people who not only can do it, but *these are the only people who can*" (emphasis mine).

At the other extreme, more adventurous executives take their Gulfstreams down to Mexico or Peru to participate in ayahuasca ceremonies with an indigenous shaman or New Age psychologist (or both). But even these invitations—like the one in my inbox that arrived this very morning as I write this chapter—are directed to "leaders" and "influencers" and promise a "highly curated" group in the hope of "generating the greatest increase in consciousness levels at scale in the shortest span of time."

Like college students who get high and then spend most of their altered state talking about the quality of the pot and where to get more, some of the entrepreneurs who are exposed to peak psychedelic states end up committing themselves to spreading the chemicals themselves—for the benefit of the world's psyche and their investors' returns.

There's nothing inherently wrong with people having powerful psychedelic experiences and then evangelizing these chemicals to the world, even for profit. No matter their motives, they are bringing potentially beneficial medicines to people suffering from depression and addiction, as well as providing new tools to those exploring consciousness and creativity. And it's naive to believe a peak mushroom experience would necessarily change a businessperson's basic nature. As Timothy Leary explained, the quality and outcome of one's trip is determined by the mindset one brings to it. An entrepreneur on mushrooms is just a psychedelic entrepreneur.

What's surprising, though, and very common among those who hold The Mindset, is their insistence that a psychedelic experience has changed their core programming. They believe that they are returning from Burning Man, the Amazon, or even commercial retreats like the "Ayahuasca Mastermind Programs for Business Leaders" offered by Entrepreneurs Awakening, as *different people*, bringing uniquely new solutions to mankind. From what I've seen, however, they return and do the very same things they were doing before—only with more cosmic justifications. The products and ideas they're pushing may change, but the methods and underlying dynamics they're employing to sell and profit off them remain the same. It just feels more profound. The quest for exponential growth—originally just a business axiom—becomes a guiding philosophy of existence and the key to saving the climate and the spirit of Gaia.

And in the worst cases, they return to the same exploitation, domination, and chauvinism they were doing before, only camouflaged in the rhetoric of global mindshift. Thanks to my early books on the psychedelic origins of digital culture, a lot of these guys call me for advice or to ask me to check out the companies, cultures, or communities they've developed—often using some of the ideas I've written about as their guiding principles. I'm usually put off from the get-go, but occasionally I get intrigued enough to sample what they've managed to create and see if I can offer any help.

One of them, a well-meaning network for rebooting humankind, launched itself with a set of invite-only gatherings in New York and San Francisco. A couple of men had experienced a particularly profound ayahuasca trip during which they realized it was their mission to gather the world's leaders and bring them to the next level of awareness so that they could address climate change. Under the supervision of a hired Zen monk, people shared doubts about their sexuality, business practices, and legacies.

Maybe six hours into all this, the conversation finally turned toward the purported subject of the event, saving the world.

How could this awakening group of elites now lead humanity to a greener, more cooperative future? *Lead?* Really? These were freshly minted New Agers whose entire life experience had been spent as financial advisors, brand managers, or tech investors. Now, thirty minutes into their awakened selves, they were ready to lead the revolution.

One came up with the idea of stock funds, filtered for bad activities like oil drilling or cigarette manufacturing, seemingly oblivious to the fact that Calvert and Ariel Investments have been doing this since the 1970s, and that everyone from Nuveen to Blackrock now offers socially responsible investment portfolios. Another thought to engage the world's youth in a more hip, media-savvy approach to climate change. I suggested she instead support Extinction Rebellion (XR), who were camped out on London's bridges at that very moment, or the Sunrise Movement, which was planning a protest just a few blocks away. I told everyone about the Post-Carbon Institute and EarthRights International, which were already producing actionable plans and policy recommendations.

"If they're so good," someone asked, "why haven't I heard of them?"

There's The Mindset, again. Why support an initiative already in progress when you can cut the ribbon on a new enterprise? As any platform enthusiast following the Web 2.0 philosophy might ask, why just work on the world's problems when you can build the WeWork for others to do it?

One multimillionaire psychedelic convert did just that. I was summoned to meet Colombian real estate mogul Rodrigo Niño shortly after he opened the Assemblage, a novel co-working space in Manhattan for spiritually inclined, psychedelic entrepreneurs. As I arrived at the converted office building, I was almost involuntarily drawn in by the inviting scents of sandalwood, patchouli, and a home-made Ayurvedic buffet just being set out by a crew of attractive young culinary artists. In fact, everyone in this place was attractive. It was like a cleaned-up Burning Man or a Phish show at Esalen.

"Rodrigo will meet you in the meditation lounge," a young woman in a white gauze dress told me as she escorted me up the wooden stairs, past a plants wall in a "biophilic design" that looked something like crop circles. She deposited me in a beautiful room with mats and pillows and a complicated device along a whole wall that played gongs and wood blocks automatically, in one of many pre-programmed meditation soundscapes.

Rodrigo eventually arrived, dismissed the young woman with a wave of his hand, sat cross-legged yet straight-backed, and told me his story. Seven years ago, he was diagnosed with stage-three melanoma, underwent a number of medical procedures, and still ended up with a poor prognosis. His money couldn't save him. "I had no choice but to venture into the unknown," he said. He went to the jungles of Peru where he took ayahuasca, and he experienced the interconnectedness of all living things as well as the presence of his own soul—something "we all have forgotten about." He went on like this, generalizing his particular experience to that of everyone else in the world.

During his ayahuasca ceremony he realized we were all still trapped, as he was, in our more limited understanding of the self. So he created this place—the Assemblage—where people could get in touch with the higher expressions of themselves.

As the sales brochure explained, the Assemblage was meant to serve entrepreneurs "at the forefront of technology, consciousness, and capital." In order to "foster transformation for the future of humanity," members of the community would learn to "elevate yourself, and elevate your business." But this was something of a bait and switch, Rodrigo told me. There was a "higher purpose, still . . ."

He wanted my help with a game. As best I could understand it, he had a vision where he saw that each of us has a second, higher self. He was in touch with his higher self, and had even given it a name. He asked me to name my own higher self. I picked the nickname my high school Spanish teacher gave me, Diego.

"Good, Diego," he said. "That would be your name in the game. That's who you would be here. Everyone here would get a name, and it would be like a fantasy role-playing adventure, except in real life. And we would use the blockchain to record everything everyone's higher self did for the community and everyone else. And when you reach certain goals, you unlock new privileges."

"So you would gamify this place?" I asked.

"First, yes. But then the game would become the reality. You would become your higher self. Get it?"

The Assemblage became a big deal, at least for a while. Deepak Chopra showed up and appeared on Rodrigo's podcast to talk about cancer, mortality, and freedom. But as it turns out, the project was principally about funneling capital from Colombia into New York real estate investments. Rodrigo eventually succumbed to his cancer, leaving thousands of angry investors, lawsuits, and charges of fraud in his wake.

The Assemblage itself closed but then reopened under new management, and Rodrigo's "game of life" was eventually released as an online experience called Akasha. But, like my meeting with the survivalist billionaires or my debate over morality with Richard Dawkins and the new atheists, the episode epitomizes an aspect of The Mindset—in this case, its approach to making the world a better place. One's personal transformation becomes the template for the transformation of everyone and everything else, with money, at scale, in what amounts to a condescendingly gamified domination of others executed from the safety of a privileged oasis.

No one brands these "game changing" approaches better than Singularity University, Silicon Valley's most self-consciously transformational incubator, consultancy, and executive training company. As SingularityU explains in its promotional literature, they are focused exclusively on supporting solutions that use "exponential technologies" to solve "global grand challenges." They're only interested in fostering "entrepreneurs and startups that take moonshots." The crime is to think linearly, which leads only to incremental improvements. Instead, we must employ the bold thinking that "aims to make something *10 times* better," challenging the status quo. Like Elon Musk going from zero to one, or their hero Ray Kurzweil rising from human to mind clone, the solutions that save the world must be one order of magnitude above the ideas of mere mortals.

Again, it's all about leadership. By becoming a premium member of Singularity University, "entrepreneurial leaders" can learn to "envision and master the future." Individuals who are daring enough to "improve the lives of billions of people" can join the executive programs at SU, learn the "power of exponential thinking," and "leverage the power of exponential technologies to make a positive impact at planetary scale." To that end, they also started the XPRIZE competition, offering grants such as $100 million for the best solution for carbon removal. With endorsers including Richard Branson, Buzz Aldrin, Tom Hanks, and Pharrell Williams, the whole project seems prophylactically insulated against buzzkill.

SingularityU's moonshot positivism has been contagious. The

normally stodgy MacArthur Foundation adapted exponential thinking to come up with their own ridiculously oversized $100 million annual mega-prize for a single proposal that solves a critical problem of our time.

The "God game" approach to planetary salvation exposes The Mindset's faulty premise. It obligates us to catalyze an evolutionary leap, to orchestrate the equivalent of a Big Bang in order to get the whole universe to conform to the exponential intentions of our species and its most influential investors. It's a sensibility that—by virtue of its ubiquity in venture philanthropy—informs even less hubristic efforts at addressing hunger, inequality, and the environment, as if one needs a totalizing, end-to-end, universal solution capable of being summarized in a TED Talk in order to be considered worthy at all. It's what we now, disparagingly, call technosolutionism.

ReGen Villages, for example, is the brainchild of former game designer James Ehrlich, an entrepreneur-in-residence at Stanford and a teacher of "disaster resilience" for Singularity University. ReGen is a total solution for the creation of regenerative and resilient communities that are capable of producing their own organic food, sourcing clean water, and educating their young, all with renewable energy and in a circular economy. Ehrlich is getting some traction—at least, with fellow Singularitarians and some of the press—with his compelling renderings of people living in high-tech harmony with nature. They grow food in domes, live in solar-powered cottages nestled into the earth, eat fresh fruit in open community courtyards, and are surrounded by woods and animals. Or at least they will be, once Ehrlich is able to convince someone to give him the funding so he can break ground.

I met up with him near his office at Stanford. He had left game design to study organic food preparation and ended up producing a TV show, *The Hippy Gourmet*, originally broadcast from Burning Man and eventually syndicated on PBS. That's how he learned about the challenges facing America's family farms, and dedicated himself to applying his skills to addressing them.

We ate vegan wraps in a Palo Alto café as he shared his plan for spawning ReGen Villages anywhere in the world. He's taken every conceivable system into consideration, from topsoil management and effluent processing to local currencies and governance. Yet even though a lot of this is supposed to be determined from the bottom up, by the people in a particular region and based on the specific climate and natural resources, the whole idea sounds a bit more like a game of SimCity than the process for a real-world community to develop. For at its core, ReGen is what Ehrlich calls a "software stack for starting, managing, and eventually autonomously improving neighborhoods."

A "software stack" is techspeak for a collection of different components, or apps, that can be used together or independently to accomplish some bigger task. So, Ehrlich's achievement has been to study many different aspects of farming, plumbing, recycling, and so on, and then develop plug-and-play computer programs that a community can use to manage its watering, seeding, electrical systems, and so on. One of them could orchestrate the processing of effluent through mango groves to produce potable water, and another could maintain proper hydration of giant contained tubes of agricultural topsoil used to grow vegetables with a minimum of watering. Then they all use sensor data to measure their effectiveness, and feed back the results for everyone's benefit and improvement.

Assuming all the pieces work—and that's a big assumption in itself—it's a beautiful picture for an organic, techno-utopian paradise. Like the Epcot Center at Disney World, but with no need for deliveries from the outside world. Almost like a space colony. Ehrlich readily admits, "We are really looking at the full stack of life support systems—sort of Mars-style, but here on Earth." Rather than helping an existing village or neighborhood utilize more regenerative principles, the ReGen project itself must be spawned on virgin territory, from the ground up, *ex nihilo*. For Ehrlich, if he ever finds the funding, this means buying a swath of forest and then clearcutting the land he needs for the farming community nestled within it. This is the way "God games" like SimCity and Civilization always work. You start with a blank slate.

It's a hubristic claim on world-building reminiscent of Walt Disney's efforts to translate the utopian simulacres of Disneyland into a plan for a real, privately owned town called Celebration—which proved a disaster. The "new urbanist" developers behind Celebration and subsequent privately planned "communities" believed they were proving that the market yielded better neighborhoods than government planning. They loved to cite urban studies legend Jane Jacobs, who was suspicious of overly zoned neighborhoods and admired mixed-use areas like Greenwich Village, which grew naturally over decades or centuries, and as the result of many different forces—including businesses—all interacting and even competing for space.

But they missed Jacobs's real point. It was the central planning she objected to, not the participation of civic and government interests. She hated Robert Moses's overarching plans for New York not because they were mandated by the government but because they were overarching plans—and, more specifically, because they called for "slum clearance" and "urban renewal" that did not respect rights and interests of the people already living in a neighborhood.

The new urbanists recast the communitarian Jacobs as a libertarian, and her appreciation for bottom-up natural urban development as an endorsement of the free market. Entirely omitting her
call for slow, natural growth of urban districts, new urbanism now
amounts to little more than a euphemism for totally planned shopping malls with apartments over the stores. Our digitally inflected
world-builders take this a step further, leveraging their billions not
just to lobby the government for the legal claim to the future, but
to serve as evidence of their own competence. They speak as if
their success building business monopolies and interactive game
worlds has earned them the right to be master planners of humanity's future.

Problem is, we're not living on a blank slate. There are people
here. And birds and trees and rocks and bacteria we barely understand. The profound irony of clearcutting a natural forest in the
name of sustainability is lost in this model. Yes, nature is in trouble, but The Mindset's approach to addressing this collective crisis
is always to *do* something. Fix it. Hack it. Reboot it. Develop it.
Scale it. Automate it. As if doing less, or even doing nothing, were
not an option. Repairing what we have, scaling back, or even seeking incremental progress doesn't make for an exciting podcast,
online panel, or TED Talk. But neither does it require massive
capital investment, sales speeches, or "buy-in."

ReGen Villages are themselves just one possible component
of an even bigger initiative conceived by Ehrlich's friend and sup-
porter Jim Rutt. The former chairman of leading systems theory
thinktank the Santa Fe Institute, Rutt has been working on his
own reboot of the world from the bottom up, called Game B.

Game B is meant to be a "civilization level social operating system," where we go from what we currently think of as Western
civilization (the failing, self-destructive Game A we are now play-

ing) to a more self-organized, networked, decentralized, and resilient way of life. Rutt has applied his widely acknowledged expertise in complex systems and game theory to work through a myriad of issues and arrive at a new model of human organization. Instead of being dominated by corporations and nation-states, we are to live and work in small, self-sovereign, kibbutz-like collectives, each with its own governance structure but linked to the others through trade, culture, and technology. It's a systems theorist's ultimate vision for a cooperative and collaborative society, working like a fractal, on many levels of coordination at once. And, I must admit, its embrace of local determination and its responsive, bottom-up approach to change is consonant with my own hopes for a society guided by communities, cooperativism, local production, and mutual aid.

How do we get from here to there? Rutt admits we should borrow as much as possible from existing ideas that work, and only invent new stuff when absolutely necessary. The initial task is to create stories, movies, and artifacts that convey the spirit of the new game, find others who are interested, and then experiment in parallel, share results, and iterate. The important thing, it seems, is to recognize you are trying to create a new game.

This determination to leave the past behind, to turn the page and move on from Game A to Game B, works better on paper—or a Playstation—than in real life. But it has become all too typical of a culture bent on saving the future by transcending the past and somehow leaving all that legacy behind. There's no time to repent or repair. We have to just bring everyone along with us into the better future we've already worked out. Start fresh. Clean slate. New planet, ecovillage, or social operating system. There's an app for this, and it's already in our programming stack. Anyone critiquing the technofix or the white male culture from which it emerges is dismissed as a paranoid Luddite or hopelessly woke fool, stuck on the sins of the past and incapable of grasping the bigger, systemic picture of which we are all part.

Yet we need to be able to critique the increasingly dominant technocratic paradigm and its deterministic approach to progress at scale, without fear of triggering the brilliant men who are trying in good faith to deploy it. It wasn't some regressive woke hippie but President Eisenhower who first warned America of the risks posed by the tech industry. His 1961 farewell address is remembered for naming the military industrial complex, but his deeper fear was of technocracy: "Yet in holding scientific discovery in respect, as we should, we must also be alert to the equal and opposite danger that public policy could itself become the captive of a scientific–technological elite."

Feminist and racial critiques have also revealed the blind spots and biases of a tech paradigm developed by a white male elite. As Bødker and Greenbaum have shown, a white male–dominated tech industry encourages the values of independence, autonomy, and distance, ignoring more circular and interconnected forces of organization. Likewise, as social feminists have argued, technologies developed by women might reflect the priorities of more people, for "they would be based in the experience of women, whose standpoint as the non-dominant group in IT provides them with a more comprehensive view of reality because of their race, class and gender."

Employed uncritically and by a homogeneous elite, the technocratic urge leads to one of two primary outcomes. At its worst, it is abused by leaders to build a totalitarian surveillance state in which every citizen's privileges are dictated algorithmically based on the data collected about them. On the other side, a more liberal technocracy is still likely to succumb to the utilitarian biases of its technologies, unintentionally neglecting the people and things that were left out of its initial calculus.

Algorithms are only as neutral as the people who program them and the parameters they are given to improve themselves. Well-meaning efforts to use computers to make prison sentencing more fair yielded algorithms that put Black people in jail for longer than whites who had committed the same crimes. Simple algorithms that bring people the kinds of news stories they are most likely to read have wreaked havoc on our civic and political lives, leading to filter bubbles, alienation, and the unchecked proliferation of fake news. None of this was intended by the technologists who programmed these systems or the people who put their faith in these game-changing improvements over the ways we were doing things before.

Technosolutions are extremely attractive to politicians and philanthropists like Michael Bloomberg, Reid Hoffman, the Ford Foundation, or Bill Gates, who take a data-driven approach to problem solving. But the funding of technological solutions to social, medical, governmental, and other problems ends up infusing the world with the values of The Mindset—as well as making us all more dependent on the companies these philanthropists founded. Whether we're talking about a smart finance grid, biohacking, drone warfare, space colonization, or universal basic income, technosolutions are too commonly informed by the values inherent in technology itself: exponential growth, automation over human intervention, forward momentum, platformization, and a disregard for existing conditions on the ground.

As a result, most moonshots turn out to be boondoggles. MIT Media Lab founder Nicholas Negroponte's much lauded One Laptop Per Child project expected to deliver 150 million $100 laptops to children in developing countries by the end of 2007. It didn't go as planned. Many countries weren't sure how they would use computers in their classrooms, particularly when their teachers were not digitally literate themselves. They didn't accept Negroponte's if-you-build-it-they-will-come premise that all kids of all cultures are born hackers who can just figure out an easy interface for themselves. By 2009, only a few hundred thousand had been shipped.

African critics complained that issues such as HIV and malnutrition were impacting people and education more than a lack of technology. While Negroponte and his team were not profiting financially from the project, they were nonetheless criticized for pushing tech solutions at Africans "that are inappropriate for them, simply to benefit [Westerners'] own need for vanity and moral reinforcement." Children who received the laptops, meanwhile, complained that the suite of music programs could only play Western beats, and not the ones they heard in songs of their own cultures.

Following the Web 2.0 philosophy for going meta, in 2014, Napster founder and first Facebook president Sean Parker spent over $40 million on Brigade, a hub for planning civic technology projects. Instead of building civic technologies, the platform would organize and provide tools for the civic technologies of others. Brigade's engineers built some clever algorithms for matching voters with their districts and elected representatives, but no one had checked to see whether civic tech developers were in need of a centralized hub. The startup was shuttered in 2019. Likewise, the 2020 Covid-19 Global Hackathon, heavily promoted by Facebook, Microsoft, and other tech companies as a way to promote technologies that could solve the pandemic, drew nearly 20,000 submissions but managed to produce, in the words of civic tech journalist and historian Micah Sifry, "a big fat nothingburger."

The belief that we can code our way out of this mess presumes the world is made of code, and anything that isn't yet code can eventually be converted to a digital format as easily as a vinyl record can be translated to a streaming file. Once the elements of the problem have been converted into data, we can use digital technology to fix them. The problem is, anything that can't be converted into code gets left behind. This puts us all in a race to get scanned, digitized, or formatted into a language compatible with the technologies orchestrating our liberties. Even solutions to the problems of technology tend to involve bringing more technology into our lives and learning to optimize our behavior in accordance with its functioning. We conform to the reward structure of the technological environment in which we live, always making more accommodations to whichever operating system our technologies—and the billionaires behind them—demand of us.

This drive toward totalitarian technocracy is what educator and media theorist Neil Postman called technopoly, the "submission of all forms of cultural life to the sovereignty of technique and technology." While we may begin using tools for our collective benefit, we slowly remake our world around the needs of technology, such as building highways and suburbs to support the automobile, or changing school curriculums so that they work on computers. Once we've done that for long enough, we eventually find ourselves inside something like a machine—a self-determinative, autonomous system that actively eliminates all other "thought-worlds." Postman says that the gods of the technopoly are efficiency, precision, and objectivity, leaving no room at all for human values, which exist in an entirely separate and unacknowledged "moral universe."

Indeed, the technopoly is inescapable, particularly for those who live to support it and who made billions finding ways to contribute to its dominance. That's why when technopolists go to the rainforest and drink from the vision vine, they see a very particular version of "all is one," and return with a zealot's vengeance to build it into reality, at scale.

We humans end up living inside The Mindset. Getting us to submit to its values becomes their biggest challenge.

10

The Great Reset

TO SAVE THE WORLD, SAVE CAPITALISM

When I saw the bodyguards coming into the hotel lobby, I assumed they were for Al Gore, who was scheduled to speak that afternoon. But when the phalanx finally made it around the corner, I realized they were protecting not the former veep but New Age legend Deepak Chopra. Why would Chopra need a security detail, I wondered, particularly at a secluded resort in Puerto Rico?

We had all been summoned there by Nobel laureate Oscar Arias for the first meeting of the Alliance for the New Humanity, billed as "the first-ever global response to the opportunity for peaceful people to work together on humanity's common challenges." This was back in 2003—over a decade before the UN adopted its 17 Sustainable Development Goals—when the idea that "societies put too much value on competition, wealth, and individualism" still felt new and somewhat radical to the elite who had benefited from those very values.

I was identified as someone who "shared the vision of a New Humanity," and invited to attend as a member of the Honorary Council, along with a variety of peacemakers and personalities from Desmond Tutu, Marianne Williamson, and Anand Shah to Guy Oseary, Jerry Hall, and Marisa Tomei. Ricky Martin was supposed to keynote, but—like pretty much all the celebrities on the list—didn't show up. Instead, various chanters, healers, meditaters, spiritualists, and retired politicians gathered before a crowd of about three hundred paying participants to share optimistic visions of how the new humanity would someday overtake our society of violence and pollution.

As Al Gore delivered the latest iteration of his PowerPoint speech "The Earth in Balance," I strolled to the back of the room. There, spread out on folding tables, were flyers advertising even more expensive workshops in prosperity, ethical business, self-care, and spiritual enlightenment, to be taught by the other panelists at various resorts around the world. There were also flyers for Chopra's most recent book, *Golf for Enlightenment*. These spiritual teachers weren't there to forge a new movement; this was business. I marveled at how skilled they were in the art of the upsell, seamlessly weaving reference to "the remaining few spots" at their next retreats into the beginning and end of any panel discussion.

The organization had also hired a public relations company to broadcast "video news reports" every few hours—pre-packaged stories for local TV news programs to air as if they were their own work. While I'd seen VNRs produced by pharmaceutical companies to stealthily advertise new products or by the oil industry to do greenwashing, I had never seen this early form of fake news practiced by NGOs or philanthropies—particularly not one with the stated purpose of making media less violent and manipulative.

Here they were, holding an expensive conference about global peace at an exotic resort hotel where the only people of color (other than centimillionaire Chopra himself) were waiters; they were talking about sustainability while eating "baby veal" and (endangered) Chilean sea bass; they were committing to fight pollution while flying thousands of miles and drinking water out of the tiniest individual-size plastic bottles I had ever seen—all to promote ideas like singer Ricky Martin's "Buenos Dias Day," when people around the world would spread goodwill through the media, Latin American style.

Like many A-list philanthrocapitalists to follow, these would-be messengers of peace ignored how their own methods undermined their bigger goals. For just as the Alliance for the New Humanity sought to combat media manipulation with more propaganda and solve the climate crisis by wasting more jet fuel, today's leading efforts at correcting the ills of capitalism, industry, and technology seek to do so with more capitalism, industry, and technology.

In her groundbreaking book *The Shock Doctrine*, Naomi Klein exposed the way oppressive governments, corporations, and wealthy individuals intentionally foment or seize upon natural and military disasters to establish neoliberal policies, entrench particular business interests, and build gated communities. So whether it's Halliburton handling the logistics for Iraq's postwar security and infrastructure, surveillance tech companies like Palantir winning contracts after 9/11, or the prison industry gaining business whenever there's an increase in poverty and crime, those who profit off crises are incentivized to perpetuate them as well as the system that keeps this feedback loop in place. The Covid pandemic created at least nine new billionaires off vaccine profits alone—

enough wealth to fully vaccinate all people in low-income countries 1.3 times.

I don't believe that philanthrocapitalists such as Mark Zuckerberg, Elon Musk, or Bill Gates are exploiting or perpetuating our global crises with the same cynical self-interest as Halliburton approaching global unrest or the Sackler family capitalizing on opiate addiction. On the contrary, they do mean, in their own ways, to solve our many problems and maybe get some of the credit. But their unthinking acceptance of The Mindset and its underlying premises renders their solutions untenable. The worse things get, the easier it is to justify The Mindset. The more we justify The Mindset, the worse things get.

For instance, Al Gore has been perhaps America's most effective champion of solar and alternative energy. Given that fossil fuels are leading to both wars and global warming, solar panels seem like a no-brainer. So, all we have to do is get venture capitalists to invest in renewable energy technologies instead of oil companies, and all this smart money will lead us to energy independence and a carbon neutral, zero-emissions, clean, green, industrial utopia. And the investors can even get richer in the process. Win win.

The problem is, while conversion of the energy grid to solar would make a lot of money for the companies building and installing solar panels, the total carbon footprint and environmental impact may not be so much better—if at all. The sun may be a renewable energy source; solar panels are anything but. They don't grow on trees, but require the mining of aluminum, copper, and rare earth metals, already in low supply. The manufacturing of solar panels is itself an extremely energy-intensive process that involves the superheating of quartz into silicon wafers, vast quantities of water, and large quantities of toxic byproducts and runoff. The solar panels themselves begin degrading just a few years after installation, and need to be replaced every decade or two. Solar panel disposal creates a host of other toxicity and environmental problems, and as long as it remains cheaper for manufacturers to dump them as landfill, we won't be seeing a robust recycling program for them anytime soon.

There lies the most fundamental problem with Mindset-derived solutions: they only move in one direction. Like anything else inspired by empirical science, the solutions all seek to dig deeper and harness some as-yet-unleveraged aspect of nature to serve our will. Like the consumer-driven, growth-based capitalism on which The Mindset is premised, these solutions usually involve finding new resources, exploiting them, selling them, and then disposing of them so more can be mined, manufactured and sold. We are free to address our environmental challenges, as long as we get more growth in the process.

If we accept capitalism and the domination of nature as basic requirements for the human project to continue, this all makes perfect sense. Solutions must make money—more money than their predecessors—in order for anyone to be incentivized to deploy them. Growth is good. "Sustainability," on the other hand, implies an unacceptable plateau in growth and development. It means partnering with nature and scaling back instead of dominating and doubling down. That's unacceptable. We must not hold back, particularly not when things get rough. We must plow through. Just on the other side of the next hill is the answer we're looking for. Have faith in scientists, technology, and market forces. We can reach new heights.

This is the philosophy underpinning the Great Reset, a campaign launched with a website and book by World Economic Forum founder Klaus Schwab. He argues for "a better form of capitalism" that encourages big investment in the businesses and technologies that can solve climate change, global poverty, and everything in between. Announced opportunistically at the height of the Covid pandemic, the Great Reset proposes a "crisis as opportunity" model of intervention, where every pain point is really just a trigger for rolling up our sleeves, getting to work, and "building back better"—with plenty of capital investment and return on that investment.

The origins of the Great Reset may actually have less to do with sustaining the planet than sustaining capitalism. It's the culmination of a twenty-year public relations campaign that began in response to protests at the World Trade Organization conference in Seattle and the Group of Eight summit in Genoa at the turn of the century. The world was changing, and environmentalists, union leaders, immigrants, and the anti-war movement were all coming to recognize global corporatism as the central cause of many of their complaints.

Schwab responded by convening a trickle of panels at WEF's conference at Davos about global warming and poverty in the global south. Even young climate activist Greta Thunberg was invited to Davos, twice. Her admonition that the assembled world leaders, corporate chiefs, and bankers not depend on carbon offsets and as-yet-uninvented technologies to solve climate change was ignored, twice. The headline that they let her speak at all was probably all they were looking for. That's because her thesis—that the world is on fire and we must immediately transition to "real zero" emissions by reducing our actual energy expenditure—contradicts the premise of the Great Reset. Schwab and the WEF believe that slowing down would be a big mistake and that market forces, unencumbered by local or national regulations, can be applied to any problem and make investors wealthier in the process.

It's a tough sell, particularly coming from the big businesses who stand to profit from it all. But the Covid crisis gave Schwab the opportunity to reframe the first stages of the Great Reset as conscious capitalism rising to what was likely the first of the world's many coming biosecurity challenges.

Mere nation-states aren't organized or cooperative enough to handle a global infection like this. As he puts it in the book, "If no one power can enforce order, our world will suffer from a 'global order deficit.' Unless individual nations and international organizations find solutions to better cooperate at the global level, we risk entering an 'age of entropy' in which retrenchment, fragmentation, anger and parochialism will increasingly define our global landscape, making it less intelligible and more disorderly." In other words, the people at the very very top of the hierarchy must use their money and technologies to restore order.

It's as if Klaus Schwab and the Davos crew have finally accepted John Barlow's Declaration of the Independence of Cyberspace, in which he claimed that nation-states were obsolete. Only a new order, some sort of technocratic network or benevolently programmed blockchain, would be up to the challenge of coordinating humanity through the coming crises. Schwab and the traditional banking elite finally bought The Mindset and are nominating themselves to lead the systemwide reboot—and to get in on the ground floor of the twenty-first century's greatest investment opportunity.

In the short term, during the pandemic, this involves funding and centralizing vaccine production, disease monitoring, and economic recovery efforts. This work then becomes the model for future interventions to address the climate, global poverty, and the rest of the United Nations' seventeen sustainability goals. Without such central leadership by a knowledgeable, wealthy elite with the best interests of the planet in mind, we are doomed to chaos. Luckily, the WEF and its invited Davos participants believe they are up to the challenge.

Schwab wants us to trust them with this great responsibility over all of our welfare. Lifting language from books by new economics theorists such as Paul Mason, Kate Rayworth, and even myself, Schwab claims the concept of "stakeholder capitalism," which will acknowledge the interests not only of shareholders but workers and locals impacted by a company's operations. Some of what he's asking for sounds great. We are to welcome the billion or more refugees displaced by climate change, listen to scientific experts, and eat less meat. All good stuff. The ways we are to arrive at this new normal are more suspect.

First, we are to liberate capital from all regulatory encumbrances—stuff like taxation, protection for local industries, and, worst of all, nationalization. Instead of forcing corporations to address global problems or taxing their winnings to do it on a national level, we are supposed to encourage their voluntary "impact investing" and support their emerging spirit of "corporate global citizenship." Thus empowered, the planet's wealthiest leaders can make good decisions from the top down.

This subjects our future welfare to the whims of wealthy individuals who believe they know best. It doesn't lead to good outcomes. The mosquito nets sent to Zambia and Nigeria by the Bill and Melinda Gates Foundation to protect people from malaria ended up poisoning local fisheries. Instead of using them to shield their beds from insects, villagers brought them to ponds and streams to use as fish nets. The tight weave caught tiny juvenile fish, devastating the reproductive cycle. The insecticide on the nets killed everything else, rendering the water unpotable as well.

Second, we are to incentivize global corporate citizenship by letting them profit off the development of new technologies. We solve for environmentally caused disease and cancers by learning to print organs. We manage resources made scarce through over-extraction by tagging everything of value and quantifying it on the blockchain. We deploy vast arrays of real-world sensors and online surveillance algorithms to track human behavior, converting it into data so it can be modeled, predicted, and influenced. Everything is made compatible with the market. So sure, it's a more "inclusive market," in that the market is able to *include everything*.

Not even progressives complain about this part. The Green New Deal is banking on the idea that the great energy transition to come will not only save the planet but give everyone jobs. They cheer when the United States or the European Union adopts new, more ambitious goals for rapid transformation of the energy infrastructure, anxious to reach carbon neutrality before global temperatures rise beyond repairable levels. They see their main challenge as convincing American workers that it's in their own best interests to get retrained for the green revolution. This is the growing industry of tomorrow. The market's requirement for growth is not an impediment to social, economic, and environmental justice, but the way to fund and reward those who bring it all about. Energy and money for everyone.

Just ask Elon Musk. His fully electric, zero emissions vehicles (along with government subsidies and carbon credits) have made him the (sometimes) richest man in the world, created jobs for over 70,000 employees, and made electric cars cool. But are Teslas really making the world a better place? They're fun to drive and a great advertisement for a post-carbon future where you can go zero to sixty in less than three seconds, but as far as environmental impact goes, they're a bit like solar panels. Although they don't spit carbon fumes while we drive them, their lifetime carbon footprint may not be much better than their gas counterparts—at least not until we change more of the power grid from coal to less carbon-intensive processes, and commit to producing renewables without the subjugation of the global south, toxic pollution, and biodiversity loss.

Even accepting that EVs and solar panels are or will one day be more energy-efficient than coal- and gas-burning technologies, the bigger question is how fast we attempt to transition. For renewables to provide a majority of our power, we would have to increase wind and solar twenty-fold. But there are not enough rare earth metals on the planet to build such an energy system and then replace it every couple of decades. Replacing a majority of our coal and gas industries with electric ones would exhaust all of our power and resources at one time, massively increasing emissions and environmental degradation in the short run. It could also increase energy inequality, by diverting power and resources to the rebuilding of the energy sector itself. Transitioning slowly, on the other hand, as things wear out, might not create such stresses, but would take many decades to bring us to zero net emissions. Both approaches result in catastrophe.

The basic laws of physics are impossible to violate. The only real answer, the really simple one that neither philanthrocapitalists nor green technologists want to hear, is that we have to reduce our energy consumption altogether. Degrowth is the only surefire way to reduce humanity's carbon footprint. It would also give us time to transition to less energy-intensive technologies. Instead of debating whether to buy electric, gas or hybrid, just keep the car you have. Better yet, start carpooling, walking to work, working from home, or working less. Like Jimmy Carter tried to tell us during his much-ridiculed fireside chats, turn down the thermostat and wear a sweater. It's better for your sinuses, and better for everyone.

Degrowth can live alongside growth-based capitalism, but it can't support it. Proponents of the Great Reset and Green New Deal believe they've come up with some kind of Grand Unified Theory for engineering a regenerative energy economy that still delivers exponential growth to its investors. Progressives may believe that this is the only way to make the idea of environmentalism palatable to the people who must either fund or permit it. But in doing so, they give cover to those who are using climate change to justify some truly egregious forms of technosolutionist profiteering, and sometimes worse.

It's hard to report on the recent history of big tech philanthropy without drifting into conspiracy theory, but that's because the cast of characters, their ties to intelligence operators, blackmail schemes, sexual impropriety, and global ambitions are so consistent. Whether or not the worst accusations about these people are true, their frequent association and multimillion-dollar partnerships reflect a shared vision for how philanthropy should be revolutionized for the twenty-first century. The Gates and Clinton foundations were launched in 2000 and 2001, and described by *Wired* as being "at the forefront of a new era in philanthropy, in which decisions—often referred to as investments—are made with the strategic precision demanded of business and government, then painstakingly tracked to gauge their success."

On the surface, this new model of venture philanthropy was an attempt to turn charity into more of a business. Instead of pouring money into lost causes, philanthropists could invest in projects that scaled so successfully they generated returns that could be invested in other ventures. Good begetting more good. But these foundations and their initiatives also provided a philanthropic halo so that a vast network of multimillion-dollar funders could support ethically questionable research agendas and personal relation-ships. Funders, scientists, and royals end up with better excuses to stay at one of Jeffrey Epstein's properties; Israeli and U.S. intelli-gence services get surveillance assets and backdoor technologies; and the Davos elite get to explore "solutions" for their own death (transhumanism) or for global inequality (eugenics).

Do just a little reading on any of these initiatives and you see names like felons Jeffrey Epstein, Ghislaine Maxwell, and Michael Milken alongside those of royals like Princes Charles and Andrew, tech founders like Bill Gates and Paul Allen, politicians like Bill and Hillary Clinton, and mega-project science advisors like Boris Nikolic and Melanie Walker. Each name serves as a trailhead to an entitled culture of would-be philosopher kings for whom con-ventional notions of morality and equity are mere obstacles to perpetuating their own dominance. They are entrenched legacies resisting any form of fundamental change.

For this global oligarchy, green investing and the oxymoroni-cally named "venture philanthropy" simply justify new forms of territorial or even interpersonal colonialism. Anything in nature can be improved or even conserved if we first convert it into a form of property, and then exploit its value in accordance with the mar-ket. Without real ownership and conscious exploitation, this logic goes, we end up with a "tragedy of the commons," where peasants or other inferiors lay waste to something valuable.

Bill Gates has employed this logic to become the biggest pri-vate owner of farmland in the United States. From an investment perspective, it allows him to meet carbon-neutral targets for sus-tainable portfolios, serving as a counterbalance to his many tech investments. But it also gives him the chance to orchestrate better land management from above. While small farmers using low-tech or even indigenous practices already know how to maintain topsoil, rotate crops, and manage runoff, Gates is certain he can improve on all that with analytic thinking. He believes he can apply science, technology, and more venture capital to develop more productive seeds, cheaper biofuels, and more advanced farming practices. Gates operates as if by purchasing resources like land and water, those with superior intelligence and foresight can manage it on behalf of all of us—using logic and technologies the rest of us couldn't understand, anyway.

Again, this is not in itself mean-spirited or selfish so much as the product of a worldview. The Mindset itself is the limiting factor here. Bill Gates enjoyed no personal financial stake in the Covid vaccines his nonprofit foundation helped develop. Yet while he encouraged cooperation between the companies racing to develop a vaccine, he also steadfastly defended their intellectual property rights. He convinced Oxford's researchers to do an exclusive deal with AstraZeneca, for example, arguing that if Big Pharma were not given a profit motive, we were at risk of "civilizational collapse." As tech writer Cory Doctorow put it at the time, "despite his cuddly reputation as a philanthropist, Gates has always pursued the ideology that the world should be guarded over by monopolist-kings, dependent on their largesse (guided by their superhuman judgment) for progress."

The result was that the wealthiest countries got vaccinated, while the poorest ones were denied patent waivers to legally produce vaccine for themselves. Gates argued, condescendingly, that the whole question was moot since the people in these countries lacked the sophistication to produce their own vaccine, anyway. The irony here is that the new mRNA vaccines are actually much easier and cheaper to produce than traditional ones. They can be made in facilities 99 percent smaller, around 99 percent cheaper, and 1,000 percent faster than previous vaccines. The production technologies were intrinsically democratizing—a problem for those who want to build wealth through capital-intensive monopolies. Over Gates's objections, President Biden temporarily suspended drugmakers' patents. Not only was this a compassionate choice, but a self-interested one. Unvaccinated populations generate more variants, which then travel back to infect people in wealthier countries, anyway.

You can monopolize, but you can't escape.

The Mindset in the Mirror

RESISTANCE IS FUTILE

The greatest danger to the holders of The Mindset would be for us to really listen to what they're telling us and react accordingly. In the techno-utopian fantasies they share from TED stages, Davos podiums, and Silicon Valley pitch decks, we human beings are regarded as little more than iron filings flying back and forth between the magnetic poles set up by the rich and powerful, mostly in an effort to keep us from impinging on their lifestyles.

How can anyone listen to World Economic Forum founder Klaus Schwab's vision for a Great Reset without getting the heebie-jeebies? His glossy brochures and high-budget videos depict a total solution for how the world's biggest banks and corporations can employ automation to fix joblessness, mass surveillance to solve immigration, biometric tracking to ensure global health, sensor networks to upgrade agriculture, blockchains to wipe out slavery, geo-engineering to remedy climate change, and capitalism to repair the extractive damage of, well, capitalism.

Such fanciful pronouncements for a civilization-wide transformation orchestrated by technocratic billionaires don't play well in Peoria, and they undermine more legitimate efforts at addressing crises, which are never so seamlessly deployed. The suspicions they engender make us less confident, for example, in mRNA vaccine technology funded, in part, by the Gates Foundation. They don't encourage our compliance with mask mandates, especially after we were originally told not to wear them. Nor do they bolster the rationale for signing onto a climate accord that gives some unelected international commission the authority to govern what kind of fuel we put in our cars or how we heat our homes.

Moreover, even when they're functioning as intended, the solution sets imposed by the technocratic elite—true to the logic of scientism—refuse to acknowledge the human soul, irrational though it may be. People want their leadership to be more than utilitarian. As nineteenth-century journalist Walter Bagehot explained, the English constitution needed two parts: "one to excite and preserve the reverence of the population," and another to "employ that homage in the work of government." The latter is the pragmatic function of Parliament; the former is the holier role of the Crown. Where the elected government values efficiency, the Crown respects dignity. Or at least, according to Bagehot, it should. Sadly, along with his complaints about the failure of the Crown to meet its divine obligations, Bagehot's later work descended into pseudoscientific racism, positing that those of mixed race lacked the "fixed traditional sentiments" on which human nature depended.

Still, what progressives' painstakingly constructed plans for job training, climate remediation, taxation, and economic equality often fail to address are the more essential needs of people to feel recognized and heard. America's form of government, in particular, emphasizes pragmatic goals and tangible things like property as the most defensible of our liberties. The Enlightenment valued logic, reason, and evidence above all else, offering a refreshing and liberating shift away from control by the church. But taken to the extreme and implemented by neoliberal technocrats, it begins to feel totalizing and disempowering—corrosive to the way people form their sense of identity, establish a connection to purpose, and experience their participation in the greater scheme of things. Traditional government assistance or The Mindset's updated universal basic income both look good on paper; still, they are poor substitutes for the dignity of getting to run one's own small business or family farm. Such enterprises were rendered all but impossible by corporate-friendly, neoliberal policies and the monopolizing power of new technologies.

Government emphasis on job training, high-tech skills, and our general compatibility with a digital future has led schools to emphasize STEM—science, technology, engineering, and math— over the softer, squishier subjects like English, social studies, and philosophy. Education has shifted away from the liberal arts, which wrestle with those fundamental questions of purpose and dignity while also building the faculties required to think critically about media and messaging. Those skills are dangerous to leave behind.

At the university level, my peers in the humanities now feel the need to frame their work and research in the language of the social sciences. They use computers to analyze the frequency of the word "thou" in Shakespeare's plays, or to "ground" philosophical premises such as aura or meaning in statistical surveys and data analysis. All this, to make their work sound more scientific so that it appeals to government, NGO, and corporate funders, who have all signed onto the greater program's metrics for success, reducing everything and everyone's worth to their utility value.

Add to this the ever-present fear of cancelation, particularly among individuals who may still be reluctant to confront their own complicity in white privilege, sexual harassment, or gender inequality, and we get a perfect storm of resentment, disenfranchisement, and paranoia. Social justice infractions are easier to prosecute in a digital media environment, where everyone's offhand statements from a decade or two ago are indelibly recorded and retrievable for later inspection. But the perfect memory of these platforms also makes it harder for people to endorse progress—especially if new rules may recast once "acceptable" behavior in a new light. Doing the right thing or using appropriate language with regard to race, gender, or sexuality today could very easily be considered wrong tomorrow. That's the nature of progress, yet it's incompatible in a world where everything is recorded and every recording is prosecutable.

The execution of social justice takes on the scientistic qualities of Richard Dawkins's appraisal of human beings as lacking any meaningful agency at all. One's intentions don't matter, since they're just an illusion perpetrated by bigger structures of repression. There's no wiggle room for the ambiguity and mixed signals inherent in human connection. Everyone is suspect, and no one has a valid excuse. In such an environment, the elite's assertions of godlike omniscience only trigger fear and paranoia—particularly when we've already been primed for suspicion and resentment by the way social media, surveillance, the gig economy and The Mindset's many other manifestations in our culture have impacted our lives and those of our loved ones.

The much-feared angry mob is real. We see them act out in alt-right conspiracy groups online, Promise Keeper rallies in the streets, threats of violence by anti-vaxxers against local school boards, and resistance to any globally coordinated mitigation of climate change. Only it's not, as *The Crowd* author Gustave Le Bon believed, a pre-existing condition of society that needs to be tamed from above, but a direct response to that top-down, technocratic effort to control them—and everything—in the first place. As the underlying logic, technology, messaging, and remote control of The Mindset is palpable everywhere—school, work, healthcare, warfare, the environment—it's no wonder so many people are frightened and angry. But instead of pushing for an alternative to the dehumanized, misogynist, antisocial, and catastrophic biases of The Mindset, the resistance is a mirror image of The Mindset itself.

In fact, the seeds of today's most virulent resistance movements were spawned long before Donald Trump's Twitter-fueled victory over establishment candidate Hillary Clinton, on subgroups of internet forums and image boards like 4chan and Reddit. Feeling blamed for society's ills, hopelessly unemployed, sexually frustrated, yet armed with laptops and The Mindset's propensity for remote attacks, this disparate network has always been ready to rumble and came to prominence during Gamergate, a series of highly coordinated online harassments against female game designers and journalists. While these young men may have been inscrutable to the establishment, the leaders of the emergent alt-right saw its members as the foot soldiers in their digital infowar against politics as usual. Steve Bannon, the media executive and political strategist who eventually served as an advisor to Donald Trump, welcomed the new population of discontents.

Already skilled at meme creation, trolling, and pranking, the frustrated and shunned young men under Bannon's thrall would be encouraged to experience themselves as a new clan of revolutionary tech bros, righting the wrongs of the castrating, politically correct left—as well the woman at their helm. For Bannon, the answer was to foment the necessary rage among the discontented for them to tear the whole establishment down, and start again from the beginning. Like a startup investor insisting on a new new thing, Bannon compares his revolutionary philosophy to Lenin's: "He wanted to destroy the state and that's my goal too. I want to bring everything crashing down, and destroy all of today's establishment." It's destructive destruction, forcing the necessary upheaval and eschewing any possibility of incremental change.

Bannon may believe that the technocrats have put Western civilization on a downward trajectory and that only a shock to the system can reverse its decline. But the irony here is that Bannon's scorched-earth anti-technocratic dream is itself based on a fringe Silicon Valley technocratic orthodoxy called accelerationism. Originating in a 1960s science fiction novel (of course), accelerationism holds that the best way forward for humanity is to accelerate computer development, automation, and global capitalism, ultimately merging human beings with digital technology. "In Silicon Valley," according to technology historian Fred Turner, "accelerationism is part of a whole movement which is saying, we don't need politics anymore, we can get rid of 'left' and 'right,' if we just get technology right."

For Bannon, the real purpose of accelerationism is to crash the system itself: run the processors and processes of technocapitalism so fast and so hard that they break down or break apart. That's why it doesn't matter what people are told or what they believe, whether it's real news or fake news, as long as it undermines their faith in the administrative state. He is adopting the same catastrophic "event" narrative as the billionaire preppers, except instead of simply preparing for Armageddon, he is actively trying to bring it on.

In this regard, he had an ally in Peter Thiel. As Thiel biographer Max Chafkin put it, "It's a fine line between trying to take advantage of developments that are already happening and trying to push them along. And I think in Thiel's career, he's stepped into not just investing in the potential collapse of existing orders but trying to accelerate them." This would explain why, in addition to investing in New Zealand properties from which he and a "cognitive elite" of "sovereign individuals" can establish a new social order after an apocalypse, Thiel also funded far-right anti-immigrant groups in the late 2010s, supported extreme political candidates, and promoted alt-right activity online, where Bannon was trying to stir things up.

Bannon leveraged Twitter and Facebook's gamelike appeal to enlist and activate new recruits in the great war against the Deep State. The online meme wars became his onboarding strategy. For, like a cult that begins as a fun game before revealing its darker and more totalizing nature, the alt-right activity on social networks seemed lighthearted at first. This was intentional. The cartoony quality of internet memes meant that if someone ever went too far with a Nazi allusion or death threat, they could always claim it was a joke—an online prank. It's just the internet, a video game; no harm done.

But there *was* harm done. Intentionally. We all saw it in the attack on the Capitol, the collapse of faith in our electoral system, and the subsequent incapacity of our elected officials and national news services to agree on even a baseline reality.

I also saw it up close as I lost one of my best friends—let's call him Sam—to the most successful campaign of the online war game, what became known as QAnon. It started out innocently enough. More like an extended, intellectually adventurous game of "What if?"—the sort of conversation Sam and I used to have in our college dorm room after a few bong hits. What if reality is a video game we forgot we're playing? What if Stanley Kubrick faked the moon landing on a movie set? What if the HARP station really can control the weather? It was a brain game that sometimes even yielded some penetrating insights on the interplay of media, technology, and the collective psyche.

QAnon "drops"—a series of cryptic messages online sup-posedly sourced from a whistleblower somewhere in the Deep State—provoked that sort of inquiry. It was always up to readers to assemble them into prophecies, as if playing a terrific fantasy role-playing game in which a narrative is composed and then A/B tested through social media. Only the most stimulating and con-tagious elements of the story survive and replicate, eventually reaching the mainstream media as they are parroted by elected politicians.

I saw the dangerous allure of this rabbit hole, but figured my friend and I were safe as long as we kept it at arm's length and remembered this was all metaphor—all a form of collective psy-choanalysis or fan fiction using Twitter posts and television news instead of dreams or science fiction novels as the original content. Many of the posts were tongue-in-cheek memes, helping to cam-ouflage the whole project as a form of sociopolitical satire. It felt derivative of the 1960s' pranksterism of Abbie Hoffman and the Yippies "levitating" the Pentagon. Or the wild speculation of con-spiracy satirist Robert Anton Wilson and the Discordians, whose "Operation Mindfuck" was meant to destabilize the consensus narrative around the Cold War and American consumer culture.

Like me, the followers of Q saw a certain soullessness among the technocratic neoliberals—and a strong possibility that at least some of the accusations against them were valid. I assumed no one actually believed the core myth—that the Democrats and their Deep State are part of a global elite that maintains its power through child sexual abuse and ritual murder. Or that they harvest a psychedelic fluid called Adrenochrome from the children's blood and consume it to enhance their power. Jeffrey Epstein's shenanigans notwithstanding, I thought we all understood that most of this stuff about politicians wasn't literally true. Rather, Q's child abuse narrative served as a great metaphor for life in the global technocracy—how we are infantilized and shafted simultaneously by godless billionaires. The real point was that if Americans really saw how the corrupt global system operates, we would be horrified. This would be what Q people call the "Great Awakening."

But it turns out my friend Sam, once a writer I turned to for counsel on the most important issues in my life, was taking all of this literally. He'd regularly text me late at night to warn me that thousands of pederasts and politicians were about to be arrested in a massive raid by the military, and that I should stay off the streets. "It's coming. Definitely this week. Stay inside."

I can only guess at what made Sam more vulnerable than me. I think it had something to do with the fact that I was a city kid, and he was raised in the country. He identified more with the plight of rural people who had been exploited and patronized by everyone from Big Agra (who took their farms) to Big Pharma (who addicted them to Oxycontin) to Big Media (who painted them as racist rednecks).

I kept asking myself, how could someone so smart have come to join this cult, believe this stuff, and engage in these antics? But maybe I was confused because I was seeing it the wrong way. Cult members aren't usually actively angry, but pacified and complacent. After all, they've found The Truth. They're smiling, not griping or complaining that their griping has been de-platformed. No, this wasn't really a cult so much as a case of classic internet addiction. Do we ever ask, "How could someone so smart have become an addict?" No, because addiction is triggered and maintained by a whole different part of one's physical and emotional makeup. If anything, addiction enlists a person's intelligence to *maintain* the supply of drugs and fend off all efforts at intervention.

What were Sam and his cohort addicted to? It wasn't the Q myth, alt-right philosophy, or any particular narrative. They were—and still are—addicted to staying online and reading and scrolling until they get that little dopamine rush that comes from connecting one dot to another. Fauci, China, Gates, 5G, Epstein, transhumanism . . . ah! It's delightful. It makes temporary sense. And then if they post the idea, it gets a few hits and likes and comments from others, and ding ding squirt squirt . . . another hit of dopamine. And another and another. As well as an ounce of dignity for being recognized. It's as if Q were simply an expression of end-stage internet addiction. The perfect digital Skinner Box and Freudian transference mechanism all at once. An industry success story.

After Trump's loss, things got worse. Sam felt betrayed. He began staying up all night reading Twitter and following links by leading alt-right posters. He became convinced that computers had been used to change the vote—presumably by either Angela Merkel from Munich or Barack Obama himself, working digitally through a consulate in Italy. The revolt at the Capitol seemed as if it would be the climax. Although the intruders managed neither to assassinate the vice president nor to prevent the election from being certified later that evening, five people died as a result of the melee. The aftermath, however, was not one of those collective realizations of having gone too far. Rather, as of this writing, in spite of FBI, Department of Justice, and Homeland Security statements to the contrary, half of all Republicans still believe left-wing activists were responsible for the attack on the Capitol.

Finally, as Joe Biden began staffing up with rather hawkish security advisors, Sam showed up on one of my messaging apps, scolding me for having stood up for the bloodthirsty establishment in my Tweets and articles. The coming bloodshed was on me. The children of red state people I don't care about would now get maimed in unnecessary wars. I should "own that."

I felt as if consensus reality had fractured, and sucked my old friend into the abyss. Social media polarization and disinformation is a big part of it. But so is the totalizing quality of The Mindset as it determines the very landscape of our culture, economy, and society—the media environment in which we try to do our reasoning and basic cognition. This was the very takeover of humans by machines that Bannon claims to be fighting against. We either mirror The Mindset or rebel in a way that reaffirms it.

The gamified values of The Mindset have trickled down to people who believe they are finally taking the red pill, escaping the simulation of the Matrix and seeing the real world as it truly is. They hate Bill Gates, Jeff Bezos, and Mark Zuckerberg for their globalist ambitions, their eventual "censorship" of conspiracy theories, and their alliances with Democrats. Yet they embrace the opportunity to be just like those self-appointed masters of the universe, remaking the world as if playing a "God game" on the computer. Accepting their role as "users" of social media platforms who actually provide all the content and labor, members of the club piece together narratives out of tiny clues, actively "doing the research," and then accept the game's fan-fiction-style lore as the hidden workings of the real world. In the manner of an improvisatory "yes-and" exercise or open-source experiment, everyone adds their own facts, however contradictory, into the canon.

Still, the Great Awakening they hope to ignite bears more than a passing resemblance to the tech billionaire fantasies they believe they are resisting. There's nothing incremental, no theory of change, no adaptation, and no compromise. Just enthusiastic anticipation of a cleansing apocalypse. Tear everything down and start over. True autonomy means total independence from all obligations to community and the conditions in which we live. Compromise is castration. We must only be satisfied with an infinity of choice and absolute liberty. This is our heritage, our destiny, and our inalienable right.

Cybernetic Karma

HOISTED BY THEIR OWN PETARD

When I was a kid, my dad and I would watch Wile E. Coyote together on Saturday mornings. *Looney Tunes* were mostly theatrical shorts repackaged for TV—much like TV is now repackaged for the internet—so while I was seeing all these cartoons for the first time, for my dad it was more an experience of revisiting movie-house shorts in a new media environment. I figured that's how he always knew what was going to happen.

"Just watch," he'd say as Coyote devised a new high-tech supertrap for Roadrunner. "He's going to get hoisted on his own petard."

I remember pondering the phrase, wondering what a petard even was, as Roadrunner ran into the trap, ate the treat, *meepmeeped*, and zoomed off unscathed. Coyote returned to his contraption and stamped on the booby-trapped platform until it suddenly exploded, collapsed on his head, and smashed him into an accordion.

Of course, it wasn't that my dad had remembered this specific episode from his youth, but that he recognized the setup. Coyote believes his superior intellect, access to technology, and place in the evolutionary order will grant him an inevitable victory over the speedy but stupid bird. Each time one of his schemes fails, he builds a bigger and more involved trap that never works as anticipated. Instead, adding insult to injury, it backfires in an even more spectacularly painful way than Coyote even envisioned.

Coyote's own hubris is always the instrument of his undoing. It's a pattern easy enough for a six-year-old to grasp. Yet it's a lesson that holders of The Mindset seem incapable of learning. No matter how smart they are, how superior to their prey, how technologically advanced, how well funded, and how preemptively insulated, they are fooling themselves if they think they're safe. None of us can escape the repercussions of our actions forever. We are all, eventually, hoisted on our own petards. While the notion of karmic retribution or a tragic flaw is ancient, the role it plays in a digital age may be unique.

It's not the technology they use but the will to conquer—the striving itself—that creates the core problem. Technology, for the most part, has simply served as a way to leverage one's advantage, or speed up the conquest. Chariots were like ancient armored vehicles, facilitating the conquest of people who hadn't even developed metallurgy. Assembly lines gave early chartered monopolies a way to hire cheap laborers by the hour, disenfranchising independent craftspeople and undermining their guilds. Gunpowder and cannons, steam engines and petrol-powered tanks all accelerated the conquest of peoples and their places, as colonialists, conquistadors, and capitalists all strove to remake their world in their own image. Without these technologies, we might never have made any of the magnificent progress in medicine, architecture, transportation, fabrication, manufacturing, agriculture, and more, about which our civilization can be both rightly proud yet also penitent for the many unintended consequences.

Until now, the mere effort of striving and forward motion has been enough to help many of the world's most aggressive conquerors and capitalists avoid the negative effects of their own activities. They "move fast" when they "break things" so they're not hit by the falling debris. Similarly, the real race to space, wealth, and whatever the tech titans think they mean by "sovereignty" is less a running toward some vision of techno-utopia than a running away from all the damage and resentment they're trying to leave behind.

Except today's unwittingly self-annihilating billionaires are a bit like Coyote in the last scene of the cartoon, where he has built his super-complex female robot to lure Roadrunner off a cliff that has been cleverly camouflaged as a safe haven. Roadrunner somehow, mysteriously, evades the trap. Coyote, seemingly oblivious to the design of his own making, careens through the artifice until he has run all the way off the cliff. He hangs there, suspended in midair until he realizes what he has done. Only then does he fall, all the way down to his fate at the bottom.

Our billionaires are in that suspended moment—as if they have driven their Teslas right off the cliffs of the Pacific Coast Highway. Looking down but not yet falling, they are hoping that some new level of striving, some next-generation technology, will allow them to squeeze out another century of progress and save them from having to suffer the inevitable comeuppance. There is nowhere left to go.

They have run out of alternatives because The Mindset only drives in a straight line toward growth and progress. Investors may call markets "cyclical," but that's really just their way of justifying pump-and-dump "shake outs" of amateur traders. For the real investor class, the trend is always up, forward, and linear. Yet their innovation and entrepreneurship are no more about making any real progress than they are about outrunning the externalities, evading the suffering, and escaping the pitchforks. Each new technological stride—from agriculture, arithmetic, and writing through to steam engines, television, and satellites—has given them the ability to evade the fallout or control our responses to it. Anything to avoid the circle, the comeuppance. Don't look back. Grow exponentially. Level up.

But digital technology may be different.

While The Mindset may have brought us into the digital age, it has not yet figured out how to contend with the primary novelty that digital has wrought: cybernetics—the circular loops generated by computers, surveillance, feedback, and interaction. The term was invented by mathematician and technology philosopher Norbert Wiener when he was designing gun mounts and radar antennas during World War II. The idea was for these systems to respond to incoming sensor data in order to adjust themselves. Instead of merely following an initial command for where to point, the gun would use feedback from its environment to find and follow its target.

Cybernetics could be used to engineer or even just understand existing mechanical systems differently, and to create things that acted more like robots. So, instead of measuring the distance between all the floors of an apartment building, an elevator uses the feedback it receives from a sensor to determine when it has reached the right place to open its doors. Likewise, the thermostat in your home "feels" when a certain temperature has been reached and turns off the heat. The sensor receives feedback from the environment, which it then "iterates" into its next decision. This changes the environment again, and so on; the thermostat joins the heater and the room into a simple, circular, self-regulating system. It can respond to changes in the environment without being given any new commands from a human controller.

This idea of feedback loops and circular systems excited a lot of people. Gregory Bateson and Margaret Mead thought cybernetics would prevent fascism. For where fascism depends on the fragmenting of knowledge and oversimplified direction from above, cybernetics would engender holism and serve as "a kind of vaccination against fragmentation." Communications experts realized they weren't just sending messages down to the masses; the masses were all reacting, speaking to one another, and changing their behaviors based on a myriad of factors. Social scientists began to look at the economy and markets as iterative systems. Even the weather, the environment, and human society itself began to make sense when they were understood as living, responding, and iterating systems, ruled only by what became known as "complexity." Everybody and everything was sensing, responding, and feeding to everyone and everything else. Linear, command-and-control logic gave way to the cycles of systems theory.

Computational machines came out of this same insight. Instead of using top-down command and control, computers accomplish their tasks through algorithms—cycles that keep repeating, like a loop, until they get their answer. Loops and loops and loops. The power of a processor is measured in terms of how many cycles it can pass through in a given second, just as the quality of a digital recording is measured in terms of its "sampling rate." Computers gave rise to chaos math and fractals, both the products of these iterative, circular processes in which the results are "fed back" into the beginning of the equation. Like a microphone "listening" to its own sound from the speaker. Feedback.

The intriguing thing about all this chaos math is that it is non-linear. Unlike the arithmetic and Euclidean geometry most of us learned in school, this stuff wasn't smooth and oversimplified. Gone were the idealized, perfect forms of ancient Greece, replaced by rough, computer-generated topographies that looked more like clouds, coral reefs, or forest floors. Somehow, out of all these super high-tech cycles, were emerging the forms of nature. Systems theory, feedback and iteration, allowed mathematicians to engage with the complexity of reality itself, rather than just sweeping all that weird stuff under the rug of approximation. Since these systems—the systems of the real world—were now understood to be nonlinear, it meant that change could come from anywhere. That's where we got the now clichéd notion that a butterfly flapping its wings in Brazil can set off a cascade of events that eventually spurs a hurricane in Texas.

These same new rules seemed to apply to human society as well. With the advent of digital technologies, the traditional, linear modes of control and communication gave way to something else. Each of us was a potential butterfly or "remote high leverage point" from which massive systemwide effects could be initiated. A kid could come up with a program that reached millions of people. A private citizen armed with a camcorder could videotape one Black man getting beaten by police and change the whole conversation about race and policing in America.

Before we had even stopped cheering for the power of small actors in giant, networked systems, however, a few guys armed with nothing but box cutters managed to crash planes into the Pentagon and the World Trade Center. With a budget of just a few thousand dollars, they turned a multitrillion-dollar transportation network against a multitrillion-dollar financial network. The darker side of remote high leverage points had revealed itself. Networks and complexity make us all vulnerable to feedback. All of our connectivity and open systems actually made us less safe.

The quest for omnipotence through technology has reached its point of diminishing returns. Not just that, but this very effort is becoming its own undoing. The internet—perhaps the ultimate accomplishment of the technocracy—is also the greatest feedback mechanism of all time. Its architects were actually clever enough to try to prevent this. The original vision of the net described by Ted Nelson was for a network of two-way links, where "anyone may publish connected comments to any page." It would have been one big dynamical system, where every link *to* something is also a link *back*. Less like a publishing platform than a nervous system. Such a system was both too difficult and, arguably, too democratic to be developed or to succeed in a media environment still largely based on one-way broadcast, so it was shelved in favor of what we now call the Web.

But the net's cyclical nature eventually shone through anyway, unleashing its cybernetic effect on everyone. If The Mindset can be understood as a unidirectional arrow of unbridled intention—westward progress, hero's journey, male climax, the eschaton—then cybernetics can be seen as the resurgence of the cyclical rhythms of nature. Indeed, nature's last laugh is that The Mindset could very well be generating its own opposite. And just in time.

We are living in the midst of a myriad of feedback loops, making it tough to figure out who is doing what to whom. We are the parents of our technologies, but we are also its users and respondents. As John Culkin, one of the fathers of media theory, explained in an article about Marshall McLuhan, "We become what we behold.

We shape our tools and then our tools shape us." While that's easy to see with straightforward technologies, such as automobiles which shaped suburban living, it's a little trickier with cybernetic ones. Each cycle is another feedback loop in which both we and our machines change, iterate, and adjust. Our algorithms are moving targets, learning new attacks whenever we develop new defense mechanisms. No one is truly in charge of where it all goes.

This has become a nightmare for those who still seek to control public sentiment. I was once called in to help a public relations firm doing "rapid response" to a corporate scandal. They had heard about my book *Present Shock* (nobody actually reads books, anymore, except you) and figured I could help them develop and operate a one-stop emergency response dashboard through which corporations in trouble could monitor "the memetic landscape." So, if a company gets caught with rat hairs in its cookies, a sweatshop in its supply chain, or sexual impropriety in its boardroom, its bosses could lock themselves up in the PR firm's situation room and enact countermeasures.

"Wait a minute," I remember the brand manager interjecting as the team read him some of the latest tweets coming through about his product. "Are these responses to our latest post, or did we just make that post in response to these tweets?" A junior executive (who actually understood how Twitter works) began reviewing the time stamps on various posts, while simultaneously explaining that if a tweet isn't a direct reply to another tweet, it's hard to know who saw what before making their own post. The brand manager was getting more anxious. The senior PR person introduced me as "the PhD" on whose theories the platform was based, and asked me to explain what was going on.

"It's a complex, dynamical system at this point," I offered. "It doesn't matter who is doing what to whom. You're all in this thing together, now." I wasn't asked to come back. My name was taken off the crisis dashboard service and I never saw a dime. But I got an inkling of what was coming. Like that screech of a microphone pointed at its own speaker, the people and processes that science and tech were invented to repress have returned as uncontrollable feedback.

Our cybernetic landscape is composed of feedback loops. Everything comes back, like karma. And though for a while it looked like digital technology was just going to accelerate the relentless drive toward infinite wealth for the few, feedback has finally kicked in, and it's not just random noise.

Just look what it has done to finance. Investors went online with a fury, and, as we've seen, used digital technology to "go meta" on the stock markets themselves. What they failed to realize, however, is that media environments tend to determine a whole lot more about the way things function inside them than we like to believe. Most hedge fund billionaires I've met really don't even make decisions anymore, beyond who to hire to write their algorithms.

So, they were particularly vulnerable when ordinary people decided to come and play in this world alongside them. I wrote my dissertation about how the stock market was becoming more like a video game ten years ago, when it still looked like digital trading platforms would remain one step ahead of human players. The discount brokers built online platforms that simulated the look and feel of those screens that professional brokers use, encouraging retail customers to day-trade and play with options contracts well beyond their skill level. As studies showed, the more frequently retail traders transacted, the more money they lost—and the more fees the platforms collected.

Discouragingly, it appeared that the traditional players would maintain their stranglehold over the economy, crushing businesses at will, with no regard for employees, small investors, or the rest of the on-the-ground economy. The bailout of the most nefarious actors behind the 2008 recession seemed to confirm our helplessness—our supposed digital empowerment notwithstanding. But all this sharing of technology and information with ground-level consumers eventually came back to haunt the big firms feeding off our human ignorance and the latency of our inferior internet connections. The gamer community analyzed the whole situation from their own perspective, and found a way to play. Like digital karma, they leveraged the power of cybernetic feedback to wage a war against big finance.

It all started on a Reddit forum called Wall Street Lulz. Someone had noticed that hedge funds had become more ruthless than ever during the Covid pandemic, shorting (betting against) the stocks of struggling retail companies in order to hasten their decline and make money off their failure. For some of these companies, like this community's cherished but declining video game store GameStop, there was actually more short interest than there were shares. These hedge funders were so sure the company would fail—or could be made to fail—that they didn't even worry about how they would cover their bets if the stock didn't tank.

So the kids on Reddit chose Gamestop as their first "meme stock" and used new, highly accessible trading platforms like Robinhood to buy as much as they could. All the gamers had to do was purchase enough shares and then hold them so that the billionaires couldn't cover their bets. The stock shot upwards, and resulting losses for those who bet against the company were incredible. To the prankster-activists, this alone was worth the cost. Then they did the same thing for AMC theaters and other favorite businesses targeted by the shorts.

Their greatest advantage was that they were not in this for

the money, but for the fun, or what they called the "lulz." This made their actions indecipherable to algorithms and the billion-aires behind them alike. The Reddit community cared less about making a profit than taking down the hedge fund billionaires who were killing vulnerable companies for a quick profit—some of which could have survived were Wall Street not using financializa-tion as a weapon against them. The financiers had abstracted the marketplace so many times that they had reduced real-world com-pany stocks not just to derivatives of derivatives, but to memes. And memes are in no one's control.

Amazingly, the response of market-makers running ultrafast trading platforms that usually benefit from lightning speed was to try to *slow things down.* As if for altruistic reasons, they argued that the kids betting on Gamestop didn't understand how the market works, and needed to be protected from their own bad judgment. But the real reason the markets needed to be slowed down was that the market-makers' true customers—the billionaire hedge funds—were getting trounced. They didn't know how to trade, not really. They were just riding the chaotic wave of their ultra-fast trading algorithms, cycling in a system they believed was funda-mentally rigged to their own advantage. They were running on automatic, which is what rendered them so vulnerable to the feed-back that was eventually generated by the system, in the form of some clever gamers on Reddit. The gamers found what hackers would call an "exploit," and the traders were hoisted by their own petard, at least for a time.

Technologies that were developed in large part to control human beings have instead turned out to be unleashing all sorts of chaotic energies. A platform like TikTok, for example, is at the very bleeding edge of persuasive technology design, com-plete with algorithmic content selection, mimetic entrainment, and surveillance features developed in China. Yet K-pop fans and other teenage prankster-activists used TikTok to organize a stunt where they ordered over a million tickets to a Trump rally—and didn't show up. As one of the organizers explained to the *New York Times*, "They all know the algorithms and how they can boost videos to get where they want . . . The majority of people who made them deleted them after the first day because they didn't want the Trump campaign to catch wind. These kids are smart and they thought of everything."

Google is also getting rocked in ways that its company philosophy, structure, and technology were designed to prevent. By surveilling employee activity, for example, Google can more easily recognize early signs of dissatisfaction or efforts at unionization. By spreading its workforce across the globe, the company makes it harder for employees to organize. Despite all this—or, more likely, because of all this—a small but growing minority of Google engineers and other workers finally formed a union in 2021.

In trying to face down the union, Google's "director of people operations" (a failed euphemism for human resources if ever there was one) made a predictable argument for technology's ability to solve labor problems by allowing the company to engage "directly with all our employees"—much in the fashion that Amazon engages "directly" with its customers, each individually. What the technologists working at the very heart of the world's biggest tech companies realize, however, is just how disempowering such individuation can be. They are the ones programming the platforms that do this to us, so they're more than aware of what it means for themselves. As one member of the union's executive council explained, Google's own actions have generated this feedback. As the engineer explains this irony, "sometimes the boss is the best organizer."

Feedback doesn't always take the form of an oppressed group

leveraging their knowledge of technology to fight back against those in power. Sometimes, it's the technology itself seeming to generate effects that work against its original purpose or those of the cultures that spawned it. The soul in the machine.

Augmented reality, for instance, the technology that allows gamers to "see" Pokemon characters when they point their smartphone cameras at various locations, is being touted by the tech industry as the next great frontier for marketing. It's the basis for Mark Zuckerberg's Metaverse. By superimposing data and graphics on streets, stores, and even merchandise, marketers can inform customers, steer them in the right direction, call attention to special sales, and create "brand experiences" on top of products. It's a huge business opportunity, as AR platforms can impact where people go, what they do, and what they buy.

The augmented reality filter on the dashboard of your car may show you that there's a McDonald's at the next exit without having indicated anything at all about the independently owned coffee shop at the current one. A business's paid placement on the visual landscape determines whether it even exists, as far as this new virtual skin on the world is concerned. Like Google Maps, it's a recipe for the monopoly service provider to make or break other businesses, and to exert tremendous control over our understanding and participation in reality.

But, as more optimistic tech thinkers point out, augmented reality may also reveal information that these very same businesses may have wanted to hide. With AR, we can access every review, comment, and price comparison. More importantly, AR can also archive the history of location. We can sit in a Broadway theater and look at images of all the plays that were ever performed there. Activist historians are already geotagging corporate logos so that when, for example, you point at a BP sign, you see a 3D image of the company's infamous undersea rupture in the Gulf of Mexico. AR can also contain and display the names of indigenous tribes displaced by colonists, images of the sacred sites on which an office building now stands, pictures of who was lynched in a town square, or videos of the cyclists mowed down on a particular city street. Digital never forgets, and cybernetics makes sure that everything eventually comes back.

Even if they can outrun all that, there's one force that the tech titans almost universally fear more than any other: artificial intelligence. In January 2015, when Elon Musk, Stephen Hawking, and Google's director of research, Peter Norvig, joined the founders of AI companies including DeepMind and Vicarious in signing an open letter about the frightening potential for artificial intelligence to end the human race, I wasn't sure how to react. Other than Hawking, these men were mostly industry developers and salesmen, and had histories of overstating the abilities of their technologies. Framing a conversation about AI with the existential risk it poses to humanity necessarily assumes that AI really works—that it can or will, as the statement explains, drive our cars, end disease, fight wars, and eradicate poverty. The only question left is how much autonomy AI will choose to grant us once it's inevitably in charge of everything.

I'm not so sure about all that. For the time being, AI and machine learning don't really work so well. They can beat humans at Jeopardy (most of the time) and chess (some of the time), but they have not gotten anywhere near what is called human-level artificial general intelligence, or AGI—the ability to do any task a human can do. Whether AI will develop human and superhuman abilities in the next decade, century, millennium, if ever, may matter less right now than AI's grip over the tech elite, and what this obsession tells us about The Mindset.

Holders of The Mindset appear less immediately afraid of AI technology itself than the people this technology is bound to replace. They know that Uber's autonomous vehicles, Amazon's robot T-shirt tailors, and future generations of AI lawyers, mortgage actuaries, and TV writers will put a whole lot of people out of work. Billionaire tech entrepreneur Mark Cuban says AI "scares the shit out of me"—but only because of how many workers will be displaced. "Things are getting faster, processing is getting faster, machines are starting to think," he explained on CNBC, adding ambiguously, "and either you make them think for you or they will take your place and do the thinking for you." The machine will be thinking for you either way, he seems to be suggesting. It's more a matter of who is working for whom. "If you are in a job where you have to think, you need to start paying attention because I guarantee you, your employer is trying to figure out ways to use technology and neural networks to do a lot of thinking that employees are currently doing."

The employer is still there. It's the employees—the displaced workers grabbing pitchforks and coming after those employers and the technologists—who pose the problem. As Reid Hoffman, founder of LinkedIn, put it, "Is the country going to turn against the wealthy? Is it going to turn against technological innovation? Is it going to turn into civil disorder?" The architects of the techno-utopian ideal now fear it will inspire a revolt of the mob that all this technology was originally invented to contain and control.

Others fear AI for what people may choose to do *with* it. Employees protested when Google acquired military robot maker Boston Dynamics in 2013, and the company eventually shed the asset. A few years later, four thousand Googlers signed a petition and at least a dozen resigned in protest over the company's decision to provide AI to Project Maven, a Pentagon program with the purpose of helping drones distinguish between targets, objects, and people.

Vladimir Putin told a group of students in 2017 that "the one who becomes the leader in this sphere will be the ruler of the world. . . . When one party's drones are destroyed by drones of another, it will have no other choice but to surrender." As if triggered by this, Elon Musk began a barrage of tweets in the following weeks, predicting that AI will be the cause of World War III, and that governments will be willing to seize AI from private firms "at gunpoint" if they see it as necessary. The misuse of AI by the wrong humans became one of Musk's primary talking points. As he explained during his South by Southwest keynote in 2018, "I think the danger of AI is much greater than the danger of nuclear warheads by a lot and nobody would suggest that we allow anyone to build nuclear warheads if they want. That would be insane. And mark my words, AI is far more dangerous than nukes. Far."

But the way people may choose to use AI is less frightening to technologists than what an AI may choose to do itself. As Stephen Hawking explained his justification for signing onto the 2015 open letter, "Whereas the short-term impact of AI depends on who controls it, the long-term impact depends on whether it can be controlled at all." Hawking gives voice to The Mindset's ultimate hubris: that it has created something that could go meta on *them*. "If a superior alien civilization sent us a message saying, 'We'll arrive in a few decades,' would we just reply, 'OK, call us when you get here—we'll leave the lights on?' Probably not—but this is more or less what is happening with AI." In the language of The Mindset, the tech titans are becoming the *zero*, and this new form of intelligence is transcending into the *one*—an order of magnitude greater than themselves. It's not the place in the exponential equation where these guys want to be. It's the vulnerable place they've been trying to put everyone and everything else in for all these years.

As a result, their fear of the coming retribution is palpable, and as vivid as a *Terminator* movie. I was at a small invite-only conference for "friends of" a tech industry leader, where I met the wealthy founder of a social media app who was so afraid of the coming age of AI that he was careful not to ever post anything negative about thinking machines. "We can talk about them here," the twenty-eight-year-old practically whispered to me, "but never on the record, and never *ever* online."

This young man's fear was that when the AIs do take over, they will review all of our social media posts in order to determine who among us are friendly to their interests and who must be eliminated—like the Chinese Cultural Revolution or the McCarthy hearings, except conducted by robots.

Yes, he had this insight while tripping on some sort of toad venom with a shaman. But on returning to work the next week and observing how his own company was using AI, he concluded that his vision of AIs networking themselves together into a new planetary governance structure was, to use his word, "inevitable." He warned me to be careful about the essays I post, and maybe to pepper them with some hints that I was only concerned for how people would exploit AI, not about the AI itself. Although he then admitted that this strategy was doomed to fail, since AIs would be able to discern such subterfuge by analyzing our linguistic patterns over time.

"Then wouldn't they be able to tell you hate them?" I asked. "Won't they be able to infer your real feelings about AI from the way you're *not* posting about this one subject?"

He paused. Then he spoke carefully, as if into a primitive translation machine. "It's not that I hate AI—I just *fear* them. That may not be interpreted as a threat to their interests."

The bigger the billionaire, the greater the fear, and the countermeasures. Elon Musk told a 2014 audience at MIT that by experimenting with AI, Larry Page and his friends at Google are "summoning the demon." In a now famous *Vanity Fair* account of a conversation between Elon Musk and DeepMind creator Demis Hassabis, Musk explained that one of the reasons he intended to colonize Mars was "so that we'll have a bolt-hole if AI goes rogue and turns on humanity." Similarly, Musk has been developing a neural net apparatus that can be lasered onto our brains, which would potentially allow us to compete with a superintelligent rogue AI that turns against us. Of course, most of Musk's space technologies are entirely dependent on AI, so a Mars mission may be less a means of escape than running straight into the robots' arms.

Maybe the fear of AI—this awareness of something they believe to be greater than themselves—will be enough to make holders of The Mindset less disdainful of the rest of humanity, and help them begin to see themselves as on the same team as everyone else. After all, they're not escaping from us; they're escaping from their own creations.

Their fate will ultimately depend on whether their artificial intelligences adopt the mindset of their creators.

Pattern Recognition

EVERYTHING COMES BACK

As I type these words, Jeff Bezos is making his first trip into space on his privately funded rocket ship, Blue Origin. He has reached Zero G a week after fellow billionaire Richard Branson got there in a slightly less explosively priapic fashion, his craft first hoisted up to the sky by some airplanes before setting off for higher altitudes.

Below them, on the surface, German towns that had stood since medieval times were being washed away by unprecedented rains; now-chronic wildfires in California were creating unsafe breathing conditions in New York; four million acres in formerly frozen Siberia were burning; and the Pacific Northwest—once considered a potential climate refuge—had just seen over eight hundred people and a billion marine animals die in a previously unimaginable heat wave that peaked at over 120 degrees Fahrenheit. And the pandemic was still raging.

As if acknowledging the externalities, Bezos used his press conference to speak directly to those on the world he had left behind. "I want to thank every Amazon employee and every Amazon customer because you guys paid for all this," he admitted. "Thank you from the bottom of my heart very much. It's very appreciated," he added, in the strangely passive, impersonal language of a customer service rep. The launch was like Amazon's version of the Macy's Day Parade, except instead of marching down Broadway with giant balloon characters for our kids, the company's largesse was in letting us bear witness to its founder's superhuman achievement. This was about Jeff.

Covering the Blue Origin flight from the desert, MSNBC's usually levelheaded and appropriately cynical anchor Stephanie Ruhle was beside herself, gushing like a teenager meeting Justin Bieber. Admittedly, any regular human, even a seasoned journalist, might be awed to be so close to the world's (sometimes) wealthiest man, and to witness a spectacle of this magnitude. Space flights are dramatic, and this one was handled with all the fanfare that the world's best public relations companies could muster. But to treat this brief flight as a milestone for humankind, particularly when NASA sent people all the way to the moon and back over fifty years ago, felt strange.

The Apollo missions may have been steeped in Cold War fears and American nationalism, but they were still a collective, public undertaking. Humanity was willfully pooling its resources in an effort to extend a civilizational pseudopod in a new direction. That first view of the earth from space in 1968 captured in a photograph helped launch the environmental movement. The image of the "blue marble" changed our civilization's perspective on our inter-relatedness and our mutual dependence on the fragile systems of nature. Even then, it was hard to justify.

This one, well, we participated in as the Covid-captive customers and gig workers of a monopoly retailer with one of the highest employee turnover rates in the industry. It was hardly a collective, public endeavor. Don't listen to Bezos's claim that this mission was the first step toward moving all of earth's heavy industry into space (as if this could somehow prove more efficient and less extractive or polluting). This was Bezos's personal triumph, childhood dream, and demonstration of power. That's why Stephanie Ruhle went weak at the knees. It wasn't that we got to space. It's that an *individual* got to space. He proved that we now live in a world where one person can make enough money to build a space program, and make good on the ultimate exit strategy.

Such is the would-be emperor's encounter with the cosmos. A singular triumph that sets one apart from the rest. Yet *space* is so much bigger than this. As Walter Benjamin explained in "To the Planetarium," his remarkable two-page essay on the invention of telescopes, "The ancients' intercourse with the cosmos had been different: the ecstatic trance." Writing shortly after he witnessed the applied technological horrors of World War II, he explained how "it is in this [collective] experience alone that we gain certain knowledge of what is nearest to us and what is remotest from us, and never of one without the other. This means, however, that man can be in ecstatic contact with the cosmos only communally. It is the dangerous error of modern men to regard this experience as unimportant and avoidable, and to consign it to the individual as the poetic rapture of starry nights."

At the time, Benjamin was thinking about the way that telescopes and star charts had turned space into something "out there." With

technology in hand, he argues, the temptation is less to encounter
nature than to master it—and to do so individually, rather than
engaging with it collectively. In other words, the truest, deepest
experience of space—of our relationship to the cosmos—may be
more richly accessible to a group of people dancing together in a
field than to a billionaire in a remote-controlled vehicle floating on
the Kármán line at the edge of the Earth's atmosphere. Or, as the
Grateful Dead's tour publicist Dennis McNally once reminded me
when I lost my backstage pass, "Relax, man, the real show's out
there in the crowd."

However we may slice it, The Mindset favors the extraordi-
nary achievements of wealthy individuals using technology to
set themselves apart from the common folk, control the natural
environment, and overcome the cycles of life. The Mindset pre-
fers straight lines, linear progress, and infinite expansion over the
ebbs and flows of the real world. Holders of The Mindset would
rather break new ground, change state, or reach a singularity than
succumb to the inevitable, compensatory undertow of natural systems. So they ignore, repress, and attempt to outrun those cycles
until eventually comes the catastrophic comeuppance.

"Apocalypses are never just complete extinction, you know,"
aboriginal scholar Tyson Yunkaporta told me when I interviewed
him for my Team Human podcast, as if to reassure me and my
worried listeners about the fate of our species under The Mindset.
"My people have been through heaps of apocalypses and they're
quite survivable, as long as you're still following the patterns of the
land and the patterns of creation. As long as you're in touch with
and moving with the landscape."

It's easy to romanticize aboriginal living, and Tyson himself
calls "indigenous" a "stupid word." He says it's "inadequate because
really what we mean is human. Everything that we described as

indigenous ways of being, these are human ways of being because we're humans and we have a habitat and we're supposed to be a habitat member."

Humans living in greater harmony with the patterns of nature don't think about owning a particular parcel of land so much as sharing a "really big home," Tyson explained to me. "You've got half a dozen 'camps' that are all like different rooms in your house. And you're moving around cleaning and being in different parts in different seasons." So if you're living in a way that's consonant with the patterns of nature, you may migrate to the river in April when it's the best time to fish. Not coincidentally, it turns out the catfish themselves provide key nutrients and medicinal value for that season. Prior to that, so you don't get attacked by mosquitoes while you're fishing, you have burned the grass in the nearby plains. The seeds of certain trees, meanwhile, have been activated to sprout by the smoke of that particular grass. Such human activities are among the many symbiotic relationships on the greater landscape.

Holders of The Mindset view any coordination with these patterns as a form of submission, and attempt instead to conquer them. Tyson went with a group of students on an excursion to a beach that is eroding into the ocean and must be fortified with sandbags and concrete retaining walls to protect the buildings there. The class exercise was to design an engineering solution to the problem. One student, appearing to be noncompliant, just sat there staring at the water. When Tyson interrogated him, he answered plainly, "Well, it's all fucked." The boy explained how the levees that had been built to block the flow of water and retain the sand on the beach were actually blocking the currents that could otherwise deposit new sand down the coast. The buildings, meanwhile, were built of concrete, which is made of sand largely dredged from the ocean floor—leaving tremendous gaping holes in the seabed. "You can build all the levees you like," he told Tyson. "But those buildings are gonna go back into the sea where they came from."

The Mindset is incapable of such observations. It depends on a Western, empirical approach to science that breaks everything down into parts rather than emphasizing the connections and interactions between all these things. This may even be an artifact of Western language systems, which tend to be more noun-based than many of their counterparts. A world of things is more static, more easily understood in terms of ownership and control, self and other. Our language has enabled certain forms of industrialism and capitalism, among other systems (like slavery and domination) that rely on objectification and categories. But it has served us less well as we seek to understand whole systems, patterns, and relationships. Of course, this all cuts both ways, informing not just our modes of oppression but the linguistic traps we fall into as we try to undo them. Many of our efforts at social justice and intersectional awareness end up becoming arguments over which labels to use rather than questioning the use of labels at all.

For our purposes here, objectifying "things"—whether for scientific identification, economic ownership, or social control—decontextualizes them from the systems of which they are a part. We think of an orange as a unit of food or a product from the grocery store, rather than the fruiting of a particular tree during a particular season. And so, we now expect to be able to eat it anywhere and anytime we choose, disregarding what this requires of the topsoil, the highway system, or our bodies to digest it out of season or divorced from its environment. In spite of ample evidence on how eating local foods is better for one's personal health as well as the greater environment, many of us persist in The Mindset's illusory contention that the supply of anything is as plentiful as the cash we have to spend on it and the greed we can summon to hoard it. Amazon and FreshDirect are happy to play to this fantasy of independence and infinite supply.

The billionaire bunker is less a viable strategy for apocalypse than a metaphor for this disconnected approach to life. The lifestyle it suggests bears more resemblance to a private, defended fortress than a welcoming oasis, because even the billionaires are aware that they've been sustaining their businesses and lifestyles on borrowed time and borrowed money. They know the edifices they've constructed are about to be swept back into the ocean.

I have borne witness to their preparations. I've been in the room as they've discussed the coming crises. I've listened as CEOs, billionaires, technologists, United Nations delegates, Pentagon officials, army generals, politicians, and even a president or two struggled to confront the ultimate repercussions of life under The Mindset. Whether they're considering climate change, economic collapse, social unrest, energy policy, or food scarcity, I'm convinced they have no real idea what's going on or what to do about it. They have no more of a clue than the rest of us. Maybe less. And I'm not sure whether that should make us feel scared or emboldened.

Almost invariably, they are still determined to think up some new paradigm just in time to save everything we've already achieved. We are to do more than simply build back better; we are to think and build *forward*. Like Coyote, we come up with one more super solution, another way of barricading the beach from the water, our lungs from pollution, our topsoil from erosion, and our technocratic model of society from its comeuppance. We are to invent a new chemical, microprocessor, blockchain, genome, nanobot, or some combination of these things to see us through to the next new world. As one former secretary of state once reassured me, "we always have, and always will. There's always another Columbus."

But there's not. Bezos is no Columbus. Columbus himself wasn't even a Columbus. The great navigators of those centuries who did journey to "new" continents were not discovering places at all, but revealing the circularity of the fixed sphere on which we live. Besides, there were already people here. Exploration exposes not the infiniteness of our potential expansion, but its limits. It makes the world smaller, not bigger.

That's not in itself a problem, unless we remain singularly fixated with moving forward as if blindered like a carriage horse. Progress doesn't always have to happen in a straight line. On the contrary, our rather recent discovery of cybernetics should free us to consider the more regenerative potential of closed loops. Contrary to the way they might be perceived by a growth-addicted venture capitalist, these regenerative systems are effectively limitless if they are not overburdened in any given moment. Snow melts and replenishes the aquifer; cows eat grass while fertilizing new growth.

Extractive, linear processes such as mining for energy resources rob from the past in order to fuel the future. We consume over three billion gallons of crude oil a day, without putting anything back, if we even could. Likewise, we lend money into existence on the expectation that the economy will keep on growing, always faster than before. When we get to apparent impasses—like the ones we're facing today—we try to innovate our way through to the other side, or transcend to some new level. Eventually, this catches up with us. We've never seen a society avoid fascism when it gets to this stage of economic inequality, or a civilization avoid collapse when it has taxed its physical environment to this extent.

Can we learn from that pattern, and avoid the same fate? Can we learn to recognize and apply regenerative principles to agriculture, energy production, and economics, so that we can bring about healthier, better distributed, and more prosperous outcomes than what otherwise awaits us?

To those afflicted by The Mindset, such circular practices are tantamount to magic. Today's investors can't grasp the concept of a founder who eschews financing and instead reinvests a business's own revenues to reach profitability. Venture capitalists call this "bootstrapping" a business, named for the way the fictional character Baron von Munchausen was able to defy the laws of physics and hoist himself up by his own bootstraps. Such basic business practices, such as earning revenue in order to grow the business, defy the exponential logic of growth through extraction and financialization.

Once, while addressing a conference of German bankers and policymakers, I told the story of how a steelworkers union applied the principles of "bounded economics" to their own retirement funds. Instead of investing them in the stock market, they began investing in construction projects that hired union steelworkers. They created jobs for themselves with their assets, which also generated returns. This worked so well that they took things a step further and invested in senior housing projects for retiring steelworkers and their parents—essentially getting three forms of return on the same investment.

"Is that legal?" one of the German bankers asked incredulously.

"Yes," I replied. "This is how bounded economics works. You don't outsource your investments to the stock market. You invest in things that come back to you or your community in multiple ways."

An economist rose and introduced himself as Doctor-Professor-something. "Mr. Rushkoff," he began. "Your ideas are interesting but, I'm sorry to say, pure fantasy." Some of the others chuckled. "Can you tell me, what is your background?"

Instead of telling them about my PhD or tenured professorship in digital economics, I simply glanced at the backdrop behind me on the stage and answered, "Blue." I may have been unnecessarily snide, but I've become frustrated by this reception. So has anyone espousing basic economic sense to those so steeped in The Mindset that they've lost the ability to think outside its unidirectional logic.

The principles for building a more circular economy that isn't dependent on growth are straightforward. Keep resources and revenue recirculating through the community, and accessible to the working class. Leverage the power of mutual aid to lift up one member of the community at a time, each according to their need. Maintain independence from big employers and disinterested investors by owning businesses cooperatively with other workers.

These ideas are threatening to traditional investors because they don't depend on their investment at all. Conventional business experts always have a reason why cooperatives, mutual aid, or local credit can never work. Freeloaders will exploit the workers, they argue. "That sounds great for progressive, educated communities like Portland or Madison," one woman in a meeting at the Aspen Institute asked me, "but do you really think inner city people have the sophistication for building cooperatives?"

It turns out the "inner city" (read: Black) people she was worried about have been at this cooperative economics thing for a very long time. The more that Black people were shunned and segregated from the rest of the American society, the more they were forced to invent the kinds of circular economic and local reinvestment strategies the rest of us are discovering only now. They pooled money to buy one another out of slavery, developed mutual aid societies to pay for each other's funerals and medical crises, and—shut out of the regular banking system—built businesses from the ground up as cooperative enterprises. Because they were forced to be self-sufficient, Black co-ops and communities of mutual aid did better than their white counterparts. This stoked resentment, and led to the rampages that targeted successful Black communities like Greenwood, Oklahoma. Some of these cooperatives are still flourishing today, largely under the radar to prevent them being "regulated" away.

These more circular systems don't make sense to holders of The Mindset because there's no endgame. Instead of climaxing in an IPO, things grow to where they need to be, and then just stay there, meeting people's needs while promoting sustainable prosperity. There's no opportunity to exit, but neither is there an obligation to grow. There's no place to externalize harm, but that becomes a strong incentive to engage in practices that benefit the community instead of poisoning or impoverishing it. This, in turn, inspires innovation and efficiencies rarely achieved when companies are financialized by distant shareholders.

McLuhan predicted that in order to orient ourselves properly in the digital age, we would need to develop pattern recognition—the ability to soften our focus from the particular details of any situation in order to perceive the greater patterns. Digital feedback loops are helping us see that our media, technology, culture, economy, and natural world all have at least as much of a cyclical character as a linear one. It's not a matter of banishing linearity and progress altogether, but rather integrating it within the greater cycles that define our existence. Not a line or a circle, but a spiral, with history never quite repeating but almost always rhyming as it moves forward through time.

With this greater understanding of the patterns underlying our past comes a greater sense of responsibility for the future. Those of us who recognize that we've been here before are the ones who have to call attention to where we are heading. Today, that means acting as a counterculture to The Mindset, introducing circularity where they see only arrows, and more thoughtful, long-term thinking when they can only strive for escape velocity.

I'm not going to offer a plan for how to save the world here, but I can point to some of what we need to do to mitigate the effects of these guys' machinations, and develop some alternative approaches. No, we don't have to ride them out on a rail. It would be too hard to draw the line between who of us are on which side. We've all participated in The Mindset, even if it was only to believe in the inevitability of our own victimhood. That's why our first step toward liberation from The Mindset is to realize that nothing is inevitable. We are not yet over the cliff. We still have choices.

Most simply, we can stop supporting their companies and the way of life that they're pushing. We can actually do less, consume less, and travel less—and make ourselves happier and less stressed in the process. Buy local, engage in mutual aid, and support cooperatives. Use monopoly law to break up anticompetitive behemoths, environmental regulation to limit waste, and organized labor to promote the rights of gig workers. Reverse tax policy so that those receiving passive capital gains on their wealth pay higher rates than those actively working for their income.

Such measures will slow or even reverse the growth rates of our largest companies, and challenge the financialized economy as it currently operates. That may go against our instinct to keep the GDP climbing, and our well-ingrained concern for the health of the economy. But since when are we humans here to serve the economy? That belief is an artifact of The Mindset, facilitated by finance, and enforced with technology.

One hedge fund manager with whom I shared these ideas told me that we have no choice but to keep growing—otherwise China will outcompete us or begin to call in the $1 trillion in U.S. debt they own. Perhaps. But right now, in spite of their country's authoritarian rule, people in China are dropping out of the rat race themselves. In response to harsh working conditions and structural inequality, many young Chinese are engaging in *tang ping*, or "lying flat" in public places as a form of leisure and protest. Instead of striving for higher pay and social status (as measured by the country's social media platforms), young people are simply lying down and making a bare minimum of effort to be productive. As Xiang Biao, a professor of social anthropology at Oxford, explained, "Young people feel a kind of pressure they cannot explain and they feel that promises were broken. People realize that material betterment is no longer the single most important source of meaning in life."

Amazingly, even if everyone does a whole lot less, we still have more than enough food and energy to go around. We'd actually have more of it. In her well-regarded paper "Beyond Growth," Gaya Herrington, a sustainability analyst for the accounting giant KPMG, explained that "Amidst global slowdown and risks of depressed future growth potential from climate change, social unrest, and geopolitical instability, to name a few, responsible leaders face the possibility that growth will be limited in the future. And only a fool keeps chasing an impossibility." She shows that while pursuing continuous growth is not possible without catastrophic climate collapse, "resource scarcity has not been the challenge people thought it would be in the 70s, and population growth has not been the scare it was in the 90s." There is ample food, water, and energy for everyone. There's just not enough to satisfy economic models that depend on infinite exponential growth. Attempting to produce that much would end civilization as we know it.

In other words, everything down here on the ground could be just fine if we weren't burdened with satisfying the needs of the abstracted map we created to represent our world for the benefit of the obscenely wealthy. We are not up against the limits of our physical reality, but the limits of our digital balance sheets. We're only in crisis because the map has replaced the territory; the virtual reality matters more than the real reality. Instead of providing for our security, our financial and technological systems are now the greatest threats to our collective wellbeing.

We are not safe behind the goggles. The virtual characters with whom we simulate intimacy may be free from disease, neurosis, neediness, and even skin pores. But there are other people in the world whom we neglect at our own peril. Not because they will storm the gates, but because the very effort to escape from them is the primary cause of the threats we now face. Yes, people and nature can be scary and unpredictable. But the attempt to control them for our own benefit doesn't work—not without a corresponding commitment to ethics, compassion, and responsibility for their wellbeing. The challenge of real reality is that there's other people here. Our own wellbeing is contingent on theirs. Maybe this is the scary truth that's been driving The Mindset all along. That's why they want to win and then get away from the rest of us as quickly and completely as possible. That's why they insist we live in a spiritual vacuum.

We can still be individuals; we just need to define our sense of self a bit differently than the algorithms do. We're not individuals to be counted, surveilled, data-analyzed, and manipulated under a pretense of convenience and connectivity. We are instead individual sensing organisms, moving into deeper relationships with other people and nature. It's the opposite trip.

In the delightful closing monologue of his Broadway show *American Utopia*, David Byrne mused on what recent discoveries about the brain tell us about this journey toward true connectivity. While the millions of unused connections in our brain are pruned as we grow into adults, perhaps they get reestablished—"only now, instead of being in our heads, they are between us and other people. Who we are is, thankfully, not just here, but it extends beyond ourselves through the connections between all of us." The artist slows us down for long enough to consider who we are and what we may be doing here.

And it's more than wishful thinking. As new research into "polyvagal theory" now suggests, there is a strong neurophysiological basis for our ability to communicate, attach, and interact with others. Most simply stated, our nervous systems do not operate independently but in concert with the other nervous systems around us. It's as if we share one collective nervous system. Our physical and mental health is contingent on nurturing those connections. Leaving others behind is futile and stupid. It's as if we've come full circle—and sensibilities that the Western world with its empirical science and individual progress were meant to transcend are back in full force.

To the extent that we have any goals at all, we should not strive for The Mindset's individual achievements, discrete wins, or profitable exits, but rather seek to make more incremental progress toward collective coherence. There's no "solution" to our woes other than maintaining a softer, more open, and more responsible comportment toward one another. We can't "fix" the world, there's no "Great Awakening," and no opportunity for "exit."

There's only the process. Our theory of change, our narrative for change, is at least as important as whatever we are going for. That makes it harder to pitch than some startup's final solution, but it is also why engaging more fully with the present is our best antidote to The Mindset's obsession with winning and escape. Our bearing, our approach, and our means are more relevant than any so-called ends.

So please, join me in listening more carefully to the promises of the tech titans and billionaire investors, as well as the world leaders in their thrall. In each and every one of their grand plans, technology solutions, and great resets, there's always an "and" or a "but"—some element of profit, some temporary compromise or cruelty, some externality to be solved at a later date, or some personal safety valve for the founder alone, along with his promise to come back for us on the next trip.

That's The Mindset's great lie, to us and them both. There is no escape, and there is no later. If we're not doing it at the moment, we're not doing it at all.

Acknowledgments

There are many people I should be thanking here. I'll refrain from trying to list you all, not just because of the many I will leave out, but because I'm not sure how many of you want to be named as accomplices.

But I simply must let Aaron Gell, my terrific editor at Medium back in 2018, share in some of the blame for this book. It was Aaron who recognized that a brief aside about a talk I did for some apocalyptic billionaires should be moved up to be the lede of my article. Thanks also to Siobhan O'Connor, Damon Beres, and Evan Williams for giving my writing so much intellectual, financial, and algorithmic support.

Thanks to my editor, Tom Mayer, for reaching out and telling me there was a book in this subject, and then pushing me to make it so much better. I'm also deeply grateful to my agent, Mollie

Glick, for advising me to write not just about ideas but the people who hold them. Thanks to Allegra Huston for lightning-fast copyediting, and to Will Scarlett for getting this book to your attention.

Thanks to my students and colleagues at Queens College for your contagious sense of mission and service. Thank you, Mara Einstein and Amy Herzog, for encouraging and defending my sabbatical. Thanks to Marina Gorbis and the Institute for the Future for your wisdom and counsel. Thanks to Mark Stahlman and the Center for the Study of Digital Life for your probes and provocations.

I am indebted to Richard Barbrook and Andy Cameron, who first identified what I'm calling "The Mindset" in their 1995 essay, "The Californian Ideology." I am also grateful to Mark Dery, who foresaw much of the lunacy I've chronicled here in his 1996 book, *Escape Velocity*.

Thanks to Nora Bateson, Reverend Billy, Stephen Brent, Luke Burgis, Amber Case, Jamie Cohen, Yael Eisenstat, Frank Faranda, Nate Hagens, HC, Renee Hobbs, Brian Hughes, Xeni Jardin, Naomi Klein, Irwin Kula, Jeremy Lent, Mark Pesce, Sarah Pessin, Vicki Robin, Philip Rosedale, Rachel Rosenfelt, Micah Sifry, Suzanne Slomin, Fred Turner, Ari Wallach, Charles Yao, Tyson Yunkaporta, and David Zweig for your brilliance, friendship, and honesty.

Thanks to everyone on Team Human for your sustenance and solidarity—especially Josh Chapdelaine and Luke Robert Mason, who protect and promote me with equal measure.

Thanks most of all to my wife, Barbara, for lovingly supporting the years of talks and travel recounted here, and my daughter, Mamie, for giving me both hope and a stake in the future.

Notes

Introduction: Meet The Mindset

5   **Elon Musk colonizing Mars:** Mike Wall, "Mars Colony Would Be a Hedge against World War III, Elon Musk Says," *Space.com*, March 28, 2018, https://www.space.com/40112-elon-musk-mars-colony-world-war-3.html.

5   **Peter Thiel reversing the aging process:** Maya Kossoff, "Peter Thiel Wants to Inject Himself with Young People's Blood," *Vanity Fair*, August 1, 2016, 2021, https://www.vanityfair.com/news/2016/08/peter-thiel-wants-to-inject-himself-with-young-peoples-blood.

5   **uploading their minds:** Alexandra Richards, "Silicon Valley billionaire pays company thousands 'to be killed and have his brain digitally preserved forever,'" *Evening Standard*, March 15, 2018, https://www.standard.co.uk/news/world/silicon-valley-billionaire-pays-company-thousands-to-kill-him-and-preserve-his-brain-forever-a3790871.html.

8   **"fairer" phones:** Bas Van Abel, interview with Douglas Rushkoff, *Team Human* podcast, March 29, 2017, https://www.teamhuman.fm/episodes/ep-30-bas-van-abel-fingerprints-on-the-touchscreen.

10 **cars into space:** Joel Gunter, "Elon Musk: The Man Who Sent His Sports

Car into Space," *BBC*, February 10, 2018, https://www.bbc.com/news/science-environment-42992143.

10 **Biosphere trials:** Steve Rose, "Eight Go Mad in Arizona: How a Lockdown Experiment Went Horribly Wrong," *Guardian*, July 13, 2020, https://www.theguardian.com/film/2020/jul/13/spaceship-earth-arizona-biosphere-2-lockdown.

Chapter 1: The Insulation Equation

13 **posted an article:** Douglas Rushkoff, "Survival of the Richest: The Wealthy Are Plotting to Leave Us Behind," *Medium*, July 5, 2018, https://onezero.medium.com/survival-of-the-richest-9ef6cddd0cc1.

16 **publishes pieces:** J. C. Cole, "American Gray SWANS USA 14 Feb 2019," *Public Intelligence Blog*, May 17, 2019, https://phibetaiota.net/2019/02/jc-cole-america-first-rooted-in-small-sustainable-distributed-farms-localize-localize-localize/, accessed September 20, 2021; and J. C. Cole, "American Gray Swans—June 2021 # 1 'Only June 2021 Petroleum Events and Other Curious Happenings!,'" *Public Intelligence Blog*, July 14, 2021, https://phibetaiota.net/2021/07/jc-cole-american-gray-swans-june-2021-1-only-june-2021-petroleum-events-and-other-curious-happenings/, accessed September 20, 2021.

17 **Rising S Company:** Rising S Company, "All Steel Bunkers and Bomb Shelters," https://risingsbunkers.com, accessed June 30, 2021.

18 **luxury underground apartments:** Elizabeth Stamp, "Billionaire Bunkers: How the 1% are preparing for the apocalypse," *CNN*, August 7, 2019, https://www.cnn.com/style/article/doomsday-luxury-bunkers/index.html.

18 **the Oppidum:** Jim Dobson, "Inside the World's Largest Private Apocalypse Shelter, The Oppidum (New Images)," *Forbes*, November 5, 2015, https://www.forbes.com/sites/jimdobson/2015/11/05/billionaire-bunker-inside-the-worlds-largest-private-apocalypse-shelter-the-oppidum/?sh=675141ef6ad6.

19 **real estate agents specializing in private islands:** Heather Murphy, "The Island Brokers Are Overwhelmed," *New York Times*, October 9, 2020, https://www.nytimes.com/2020/10/09/realestate/private-islands-coronavirus.html.

19 **Cancer-causing microplastics:** Jamie Wheal, *Recapture the Rapture: Rethinking God, Sex, and Death in a World That's Lost Its Mind* (New York: HarperCollins, 2021).

19 **World Wide Fund for Nature study:** Wijnand de Wit and Nathan Bigaud, "No Plastic in Nature: Assessing Plastic Ingestion from Nature to People," World Wide Fund for Nature, 2019, https://d2ouvy59p0dg6k.cloudfront.net/downloads/plastic_ingestion_web_spreads.pdf.

21 **"aquapreneurs":** Joe Quirk and Patri Friedman, *Seasteading: How Floating Nations Will Restore the Environment, Enrich the Poor, Cure the Sick, and Liberate Humanity from Politicians* (New York: Free Press, 2017).

21 **humanity's return to the sea:** "Busan UN Habitat and OCEANIX Set to Build the World's First Sustainable Floating City Prototype as Sea Levels Rise," UNHabitat.org, November 18, 2021, https://unhabitat.org/busan-un-habitat -and-oceanix-set-to-build-the-worlds-first-sustainable-floating-city-prototype -as.

21 **"To establish permanent":** https://www.seasteading.org.

21 **"We've had the agricultural revolution":** https://www.seasteading.org.

Chapter 2: Mergers and Acquisitions

25 **Tech companies actively sought:** Douglas Rushkoff, *Cyberia: Life in the Trenches of Hyperspace* (New York: HarperOne, 1994).

25 **"new communalists":** Fred Turner, *From Counterculture to Cyberculture: Stewart Brand, the Whole Earth Network, and the Rise of Digital Utopianism* (Chicago: University of Chicago Press, 2006).

26 **Operation Sundevil:** Bruce Sterling, *The Hacker Crackdown: Law and Disorder on the Electronic Frontier* (New York: Bantam, 1992).

26 **"Governments of the Industrial World":** John Perry Barlow, "A Declaration of the Independence of Cyberspace," Electronic Frontier Foundation, 1996, https://www.eff.org/cyberspace-independence.

26 **fungus and bacteria:** Qi Hui Sam, Matthew Wook Chang, and Louis Yi Ann Chai, "The Fungal Mycobiome and Its Interaction with Gut Bacteria in the Host," *International Journal of Molecular Sciences*, February 4, 2017, https://www .ncbi.nlm.nih.gov/pmc/articles/PMC5343866/.

28 **extolled the virtues of the deal:** Saul Hansell, "America Online Agrees to Buy Time Warner for $165 Billion; Media Deal is Richest Merger," *New York Times*, January 11, 2000, https://www.nytimes.com/2000/01/11/business/media -megadeal-overview-america-online-agrees-buy-time-warner-for-165-billion .html.

28 **the piece I wrote placed in the** *Guardian:* Douglas Rushkoff, "Why Time Is Up for Warner," *Guardian*, January 20, 2000, https://www.theguardian.com/ technology/2000/jan/20/onlinesupplement10.

29 **People blamed:** Seth Stevenson, "The Believer," *New York Magazine*, July 6, 2007, https://nymag.com/news/features/34454/.

30 **hired investment bank Salomon Smith Barney:** Tim Arango, "How the AOL–Time Warner Merger Went So Wrong," *New York Times*, January 10, 2010, https://www.nytimes.com/2010/01/11/business/media/11merger.html.

31 **probably borrowed:** Steven Levy, *Facebook: The Inside Story* (New York: Blue Rider Press, 2020).

32 **stocks quadruple:** Lisa Pham, "This Company Added the Word 'Blockchain' to Its Name and Saw Its Shares Surge 394%," *Bloomberg*, October 27, 2017, https://www.bloomberg.com/news/articles/2017-10-27/what-s-in-a-name-u-k-stock-surges-394-on-blockchain-rebrand.

33 **"independent, host-led local organizations":** Dave Lee, "Airbnb Using 'Independent' Host Groups to Lobby Policymakers," *Financial Times*, March 21, 2021, https://www.ft.com/content/1afb3173-444a-47fa-99ec-554779dde236.

33 **Google was outspending:** Shaban Hamza, "Google for the First Time Outspent Every Other Company to Influence Washington in 2017," *Washington Post*, January 23, 2018, https://www.washingtonpost.com/news/the-switch/wp/2018/01/23/google-outspent-every-other-company-on-federal-lobbying-in-2017/.

33 **outspent by Facebook:** Lauren Feiner, "Facebook Spent More on Lobbying than Any Other Big Tech Company in 2020," *CNBC*, January 22, 2021, https://www.cnbc.com/2021/01/22/facebook-spent-more-on-lobbying-than-any-other-big-tech-company-in-2020.html.

33 **Numerous studies:** Martin Gilens and Benjamin I. Page, "Testing Theories of American Politics: Elites, Interest Groups, and Average Citizens," *Perspectives on Politics* 12, no. 3 (2014): 564–81, https://doi.org/10.1017/S1537592714001595; Chris Tausanovitch. "Income, Ideology, and Representation," *RSF: The Russell Sage Foundation Journal of the Social Sciences* 2, no. 7 (2016): 33–50, https://doi.org/10.7758/rsf.2016.2.7.03.

33 **"citizens and mass-based interest groups":** Martin Gilens and Benjamin I. Page, "Testing Theories of American Politics: Elites, Interest Groups, and Average Citizens," *Perspectives on Politics* 12, no. 3 (2014): 564–81, https://doi.org/10.1017/s1537592714001595.

34 **"motor resonance":** Jeremy Hogeveen, Michael Inzlicht, and Sukhvinder S. Obhi, "Power Changes How the Brain Responds to Others," *Journal of Experimental Psychology: General* 143, no. 2 (2014): 755–62, https://doi.org/10.1037/a0033477.

34 **damage to the brain's orbitofrontal lobes:** Dacher Keltner, "The Power Paradox," *Greater Good Magazine*, December 1, 2007, https://greatergood.berkeley.edu/article/item/power_paradox.

34 **"we've decided that capitalism":** For more on this, see Scott Galloway, *Post Corona: From Crisis to Opportunity* (New York: Portfolio, 2021), and the brief interview in Adam Shapiro, "Capitalism 'Will Collapse on Itself' without More Empathy and Love: Scott Galloway," *Yahoo!*, December 1, 2020, https://finance.yahoo.com/news/capitalism-will-collapse-on-itself-without-empathy-love-scott-galloway-120642769.html.

34 **Government readily bailed out:** Kumutha Ramanathan, "Former US Fed Governor Warns Global Economy Will Take a Long Time to Recover," *Yahoo! Finance,* October 23, 2020, https://finance.yahoo.com/news/former-us-fed -governor-randall-kroszner-global-markets-coronavirus-pandemic-recovery -warning-050012598.html.

Chapter 3: A Womb with a View

36 **systematically excluded:** Emma Goldberg, "Women built the tech industry. Then they were pushed out," *Washington Post,* February 19, 2019, https://www .washingtonpost.com/outlook/2019/02/19/women-built-tech-industry-then -they-were-pushed-out/.

37 **"about the quality . . . 'in your pajamas'":** Stewart Brand, *The Media Lab: Inventing the Future of M.I.T.* (New York: Viking Penguin, 1987), 251.

38 **"we're way closer . . . people's brains":** Matthew Gault, "Billionaires See VR as a Way to Avoid Radical Social Change," *Wired,* February 15, 2021, https:// www.wired.com/story/billionaires-use-vr-avoid-social-change/.

38 **"It is not possible . . . you wanted":** Gault, "Billionaires See VR as a Way to Avoid Radical Social Change."

39 **"Yes, we are in a pandemic":** David Zweig, "$25,000 Pod Schools: How Well-to-Do Children Will Weather the Pandemic," *New York Times,* July 30, 2020, https://www.nytimes.com/2020/07/30/nyregion/pod-schools-hastings-on -hudson.html.

41 **legions of Amazon workers:** Joey Hadden, "Amazon Delivery Drivers Share What It's Like to Be on the Front Lines of the Coronavirus Pandemic, Including Not Having Time to Wash Their Hands and Uncleaned Vans," *Business Insider,* April 2, 2020, https://www.businessinsider.com/why-amazon-delivery -workers-feel-exposed-and-vulnerable-to-coronavirus-2020-3.

41 **Amazon avoids taxes:** Matthew Gardner, "Amazon Has Record-Breaking Profits in 2020, Avoids $2.3 Billion in Federal Income Taxes," Institute on Taxation and Economic Policy, February 3, 2021, https://itep.org/amazon-has -record-breaking-profits-in-2020-avoids-2-3-billion-in-federal-income-taxes/.

41 **anti-competitive practices:** Mark Chandler, "Amazon Accused of Anti-Competitive Practices by US Subcommittee," *Bookseller,* October 8, 2020, https://www.thebookseller.com/news/amazon-accused-anti-competitive -practices-us-subcommittee-1222115.

41 **abuses labor:** Jodi Kantor, Karen Weise, and Grace Ashford, "Power and Peril: 5 Takeaways on Amazon's Employment Machine," *New York Times,* June 15, 2021, https://www.nytimes.com/2021/06/15/us/politics/amazon-warehouse -workers.html; Casey Newton, "Amazon's Poor Treatment of Workers Is

Catching up to It during the Coronavirus Crisis," *Verge*, April 1, 2020, https://www.theverge.com/interface/2020/4/1/21201162/amazon-delivery-delays-coronavirus-worker-strikes.

42   **more people opened online trading:** Annie Massa, "Pandemic-Fueled Day Trading Is Overwhelming Online Brokers—and the Traders Are Fuming," *Fortune*, December 9, 2020, https://fortune.com/2020/12/08/day-trading-online-brokers-tech-failure-crashes-outages/.

42   **Shares of Zoom went up:** Shanhong Liu, "Price of Zoom shares traded on Nasdaq Stock Market in 2020 and 2021," *Statista*, August 9, 2021, https://www.statista.com/statistics/1106104/stock-price-zoom/.

42   **Jeff Bezos's fortune rose:** Chase Peterson-Withorn, "How Much Money America's Billionaires Have Made During the Covid-19 Pandemic," *Forbes*, April 30, 2021, https://www.forbes.com/sites/chasewithorn/2021/04/30/american-billionaires-have-gotten-12-trillion-richer-during-the-pandemic.

42   **five biggest U.S. tech companies:** The staff of the *Wall Street Journal*, "How Big Tech Got Even Bigger," *Wall Street Journal*, February 6, 2021, https://www.wsj.com/articles/how-big-tech-got-even-bigger-11612587632.

42   **Netflix's share price:** Jonathan Ponciano, "5 Big Numbers That Show Netflix's Massive Growth Continues during the Coronavirus Pandemic," *Forbes*, October 20, 2020, https://www.forbes.com/sites/jonathanponciano/2020/10/19/netflix-earnings-5-numbers-growth-continues-during-the-coronavirus-pandemic/.

43   **"Many creatives, startups, and techies":** Jen Murphy, "Remote Workers Flee to $70,000-a-Month Resorts While Awaiting Vaccines," *Bloomberg*, February 15, 2021, https://www.bloomberg.com/news/articles/2021-02-15/remote-workers-flee-to-luxury-beach-resorts-while-awaiting-vaccines.

43   **"It's been great here":** Julie Satow, "Turning a Second Home into a Primary Home," *New York Times*, July 24, 2020, https://www.nytimes.com/2020/07/24/realestate/coronavirus-second-homes-.html.

43   **Each 1 percent increase:** Mary Van Beusekom, "Race, Income Inequality Fuel COVID Disparities in US Counties," Center for Infectious Disease Research and Policy, University of Minnesota, January 20, 2021, https://www.cidrap.umn.edu/news-perspective/2021/01/race-income-inequality-fuel-covid-disparities-us-counties; Tim F. Liao and Fernando De Maio, "Association of Social and Economic Inequality with Coronavirus DISEASE 2019 Incidence and Mortality across US Counties," *JAMA Network Open* 4, no. 1 (2021), https://doi.org/10.1001/jamanetworkopen.2020.34578.

43   **people processing pork:** Tina L. Saitone, K. Aleks Schaefer, and Daniel P. Scheitrum, "COVID-19 Morbidity and Mortality in U.S. Meatpacking Coun-

ties," *Food Policy* 101 (2021): 102072. https://doi.org/10.1016/j.foodpol.2021
.102072.

Chapter 4: The Dumbwaiter Effect

47 **"The integration of all Uber brands . . . life":** Megan Rose Dickey, "Uber
Unveils New Skyport Designs for Uber Air," *TechCrunch*, June 11, 2019, https://
techcrunch.com/2019/06/11/uber-unveils-new-skyport-designs/.

47 **RVs and those living in cars:** Marina Gorbis, "Hiding in Plain Sight: America's Working Poverty Epidemic," *Medium*, April 14, 2021, https://medium
.com/institute-for-the-future/hiding-in-plain-sight-americas-working-poverty
-epidemic-740f0b7202ea.

49 **"renewables":** Richard Maxwell and Toby Miller, *Greening the Media* (New
York: Oxford University Press, 2012).

49 **"the dumbwaiter effect":** Douglas Rushkoff, *Throwing Rocks at the Google Bus:
How Growth Became the Enemy of Prosperity* (New York: Penguin Portfolio, 2016),
19.

50 **miscarriages, cancers, and shortened lifespan:** *Producing the Fairphone*,
directed by Geert Rozinga (De Eerlijke Onderneming, 2016), https://www
.vpro.nl/programmas/tegenlicht/kijk/afleveringen/2016-2017/de-eerlijke
-onderneming.html.

51 **"create the perfect tee":** Amazon, "Made for You," https://www.amazon.com/
stores/made+for+you/page/E853E0F0-6F79-442D-B7E8-3A0E0531FAF2,
accessed August 9, 2021.

Chapter 5: Selfish Genes

53 **John Brockman:** I was a client of the John Brockman Agency beginning about
ten years after this party, from 2009 to 2019.

56 **robust sense of justice:** See Emmanuel Lévinas, *Totality and Infinity: An Essay on
Exteriority*, trans. Alphonso Lingis (Pittsburgh: Duquesne University Press, 1969).

57 **"parasitic worm":** Richard Kearney, *Anatheism: Returning to God After God*
(New York: Columbia University Press, 2009), 168–71.

57 **seed hundreds of women:** James B. Stewart, Matthew Goldstein, and Jessica Silver-Greenberg, "Jeffrey Epstein Hoped to Seed Human Race with His
DNA," *New York Times*, July 31, 2019, https://www.nytimes.com/2019/07/31/
business/jeffrey-epstein-eugenics.html.

57 **Lolita Express:** Julia La Roche, "Jeffrey Epstein Attended the 'Billionaires'
Dinner' and Now His Presence Has Been Scrubbed," *Yahoo! Finance*, July 15,

2019, https://www.yahoo.com/now/jeffery-epstein-billionaires-dinner-john
-brockman-photos-sarah-kellen-173443481.html.

58 **impregnate twenty women at a time:** Stewart, Goldstein, and Silver-Greenberg, "Jeffrey Epstein Hoped to Seed Human Race with His DNA."

58 **freeze his head and penis:** Bess Levin, "Jeffrey Epstein Wanted to Have His Penis
Frozen and 'Brought Back to Life in the Future,'" *Vanity Fair*, July 31, 2019, https://
www.vanityfair.com/news/2019/07/jeffrey-epstein-transhumanism-cryonics.

58 **attributed to:** These quotes appear in Bacon's posthumously published *Mas-
culus Partus Temporum*, (*The Masculine Birth of Time*, 1603), which some schol-
ars believe was used or even fabricated by later members of the Royal Society
to justify their own misogyny and subjugation of both women and nature.
For more on Bacon as part of science's misogynist foundations, see Carolyn
Merchant, *The Death of Nature: Women, Ecology and the Scientific Revolution* (San
Fransisco: HarperSanFrancisco, 1983). For more on how Bacon may have been
misrepresented, see Alan Soble, "In Defense of Bacon," in *A House Built on Sand:
Exposing Postmodernist Myths about Science*, ed. Noretta Koertge, (Oxford Uni-
versity Press, 1998), 195–215.

58 **"I am come in very truth . . . herself":** Clifford D. Conner, *A People's History of
Science* (New York: Nation Press, 2005), 364.

59 **"atheistical" materialism:** Charles Webster, *From Paracelsus to Newton: Magic
and the Making of Modern Science* (Cambridge: Cambridge University Press,
1982), 99–102.

61 **"survival machines":** Richard Dawkins, *The Selfish Gene: 40th Anniversary Edi-
tion* (New York: Oxford University Press, 2016), xxix.

62 **master classes in behavioral economics:** Daniel Kahneman, "A Short Course
in Thinking About Thinking,'" Edge Masterclass 2007, https://www.edge
.org/events/the-edge-master-class-2007-a-short-course-in-thinking-about
-thinking; Richard Thaler, Sendhil Mullainathan, and Daniel Kahneman, "A
Short Course in Behavioral Economics," Edge Master Class 2008, https://www
.edge.org/event/edge-master-class-2008-richard-thaler-sendhil-mullainathan
-daniel-kahneman-a-short-course-in.

Chapter 6: Pedal to the Metal

68 **"The power to dominate . . . authoritarian":** Riane Eisler, *The Chalice and the
Blade: Our History, Our Future* (New York: HarperCollins, 1987), 86.

69 **Initially the market economy:** For a description of how these money systems
worked, see Douglas Rushkoff, *Life Inc.: How Corporatism Conquered the World,
and How We Can Take It Back* (New York: Random House, 2009).

69 **People worked less:** Bernard Lietaer and Stephen M. Belgin, *Of Human Wealth: Beyond Greed and Scarcity* (unpublished manuscript, 2004).

70 **"state of nature" . . . "absolute dominion":** John Locke, *The Second Treatise on Civil Government and A Letter Concerning Toleration* (Oxford: B. Blackwell, 1948).

73 **invisible population of rich people:** Since 1975, according to a study by RAND, the wealthiest 1 percent of Americans have diverted $50 trillion from the rest of us. Carter C. Price and Kathryn A. Edwards, "Trends in Income From 1975 to 2018," RAND Corporation, 2020, https://www.rand.org/pubs/working_papers/WRA516-1html.

73 **extracting wealth from the poor:** Jon Evans, "GrubHub/Seamless's Pandemic Initiatives Are Predatory and Exploitative, and It's Time to Stop Using Them," *TechCrunch*, April 5, 2020, https://techcrunch.com/2020/04/05/its-time-to-stop-using-grubhub-seamless-forever/.

73 **This is not creative destruction:** Lachlan Carey and Amn Nasir, "Something for Nothing? How Growing Rent-Seeking Is at the Heart of America's Economic Troubles," *Journal of Public and International Affairs*, https://jpia.princeton.edu/news/something-nothing-how-growing-rent-seeking-heart-americas-economic-troubles.

74 **"some kinds of social change":** Jennifer Szalai, "Steven Pinker Wants You to Know Humanity Is Doing Fine. Just Don't Ask About Individual Humans," *New York Times*, February 28, 2018, https://www.nytimes.com/2018/02/28/books/review-enlightenment-now-steven-pinker.html.

74 *The Dawn of Everything:* David Graeber and David Wendgrow, *The Dawn of Everything: A New History of Humanity* (New York: Farrar, Straus and Giroux, 2021).

74 **problem with Pinker's oft-quoted statistics:** Jeremy Lent, "Steven Pinker's Ideas Are Fatally Flawed. These Eight Graphs Show Why," *openDemocracy*, May 21, 2018, https://www.opendemocracy.net/en/transformation/steven-pinker-s-ideas-are-fatally-flawed-these-eight-graphs-show-why/.

75 **"As we have seen":** Steven Pinker, *Enlightenment Now: The Case for Reason, Science, Humanism, and Progress* (New York: Penguin, 2018), 109.

75 **"What we want . . . fast as you can":** "Dr. Jordan Peterson Makes the Case for Capitalism," YouTube video, July 5, 2020, 10:05, https://www.youtube.com/watch?v=uWeDnN0O_xA.

76 **Extrinsic rewards . . . intrinsic rewards:** Alfie Kohn, *Punished by Rewards: The Trouble with Gold Stars, Incentive Plans, A's, Praise, and Other Bribes* (New York: Houghton Mifflin, 1993); Kenneth Thomas, "The Four Intrinsic Rewards That Drive Employee Engagement," *Ivey Business Journal*, December 4, 2017, https://iveybusinessjournal.com/publication/the-four-intrinsic-rewards-that-drive-employee-engagement/.

76 **"vile despoilers":** Pinker, *Enlightenment Now*.

76 **Bezos has a yacht:** Allison Morrow, "Jeff Bezos' Superyacht Is So Big It Needs Its Own Yacht," *CNN*, May 10, 2021, https://www.cnn.com/2021/05/10/business/jeff-bezos-yacht/index.html.

77 **"there is a growing loss":** Pope Francis, *Fratelli Tutti*, sec. 13, https://www.vatican.va/content/francesco/en/encyclicals/documents/papa-francesco_20201003_enciclica-fratelli-tutti.html.

78 **Nietzsche's sister:** Sue Prideaux, "Far Right, Misogynist, Humourless? Why Nietzsche Is Misunderstood," *Guardian*, October 6, 2018, https://www.theguardian.com/books/2018/oct/06/exploding-nietzsche-myths-need-dynamiting.

78 *übermensch* **wannabes:** Alex Ross, "Nietzsche's Eternal Return," *New Yorker*, October 4, 2019, https://www.newyorker.com/magazine/2019/10/14/nietzsches-eternal-return.

78 **"I no longer believe":** Ross, "Nietzsche's Eternal Return."

79 **"it's hard to find actual examples":** Steven Levy, "Google's Larry Page on Why Moon Shots Matter," *Wired*, January 17, 2013, https://www.wired.com/2013/01/ff-qa-larry-page/.

80 **Zuckerberg told** *The New Yorker*: Evan Osnos, "Can Mark Zuckerberg Fix Facebook before It Breaks Democracy?," *New Yorker*, September 10, 2018, https://www.newyorker.com/magazine/2018/09/17/can-mark-zuckerberg-fix-facebook-before-it-breaks-democracy.

81 **"I don't expect":** Rick Merritt, "Moore's Law Dead by 2022, Expert Says," *EE Times*, August 27, 2013, https://www.eetimes.com/moores-law-dead-by-2022-expert-says/.

Chapter 7: Exponential

83 **Goldman Sachs:** Adrian Ash, "Goldman Sachs Escaped Subprime Collapse by Selling Subprime Bonds Short," *Daily Reckoning*, October 19,2007, http://www.dailyreckoning.com.au/goldman-sachs-2/2007/10/19.

85 **"No man who":** Kenneth T. Jackson, *Crabgrass Frontier: The Suburbanization of the United States* (New York: Oxford University Press, 1985), 231.

85 **Federal Housing Authority:** C. Lowell Harriss, *History and Policies of the Home Owners' Loan Corporation* (New York: National Bureau of Economic Research, 1951), 41–48.

86 **GE eventually sold:** GE, "GE Completes the Separation of Synchrony Financial," November 17, 2015, https://www.ge.com/news/reports/ge-completes-the-separation-of-synchrony-financial.

87 **"world trade . . . speed of light":** Nicholas Negroponte, *Being Digital* (New York: Knopf, 1995).

87 **"new paradigm":** Alen Mattich, "The New 'New Paradigm' for Equities," *Wall Street Journal*, May 28, 2013, https://www.wsj.com/articles/BL-MBB-1982.

89 **"competition is for losers":** Peter Thiel, "Competition Is for Losers," *Wall Street Journal*, September 12, 2014, https://www.wsj.com/articles/peter-thiel -competition-is-for-losers-1410535536.

89 **"fidelity to an event":** Peter Thiel and Blake Masters, *Zero to One: Notes on Startups, or How to Build the Future* (New York: Crown Business, 2014). He borrows this idea from Maoist Alain Badiou.

91 **spending electricity on purposeless calculations:** Bitcoin advocates argue that some percent of this energy comes from "renewable" sources. Crypto critics argue that these figures don't even account for the additional 50 percent of energy being consumed by rival proof-of-work tokens.

92 **The companies behind our activity trackers:** A. J. Perez, "Use a Fitness App to Track Your Workouts? Your Data May Not Be as Protected as You Think," *USA Today*, August 16, 2019, https://www.usatoday.com/story/sports/2019/08/16/ what-info-do-fitness-apps-keep-share/1940916001/.

94 **"600 megabytes compressed":** Janet Lowe, *Google Speaks: Secrets of the World's Greatest Billionaire Entrepreneurs, Sergey Brin and Larry Page* (New York: Wiley, 2009), 239.

94 **"life is just bytes":** Jeremy Lent, *The Web of Meaning: Integrating Science and Traditional Wisdom to Find Our Place in the Universe* (Gabriola, BC, Canada: New Society, 2021).

94 **how a life form expresses itself:** Marc H. V. van Regenmortel, "Reductionism and complexity in molecular biology. Scientists now have the tools to unravel biological and overcome the limitations of reductionism," *EMBO Reports* 5, no. 11 (2004): 1016–20, https://doi.org/10.1038/sj.embor.7400284.

Chapter 8: Persuasive Tech

99 **"the pictures in their heads":** Walter Lippmann, *Public Opinion* (New York: Harcourt, Brace, 1921).

100 *The Crowd:* Gustave Le Bon, *The Crowd: A Study of the Popular Mind* (New York: Viking, 1960).

101 **"manufacture consent":** Lippmann, *Public Opinion.*

102 **"mechanical, advanced, and necessary":** Edward Bernays, *Propaganda* (Brooklyn, NY: Ig Publishing, 2004).

102 **the madness of crowds:** Le Bon, *The Crowd.*

103 **"technology of behavior":** B. F. Skinner, *Science and Human Behavior* (New York: Macmillan, 1953).

103 **"a servosystem coupled":** Fred Turner, *The Democratic Surround: Multimedia*

and American Liberalism from World War II to the Psychedelic Sixties (Chicago: University of Chicago Press, 2013), 123.

103 **"How would we rig":** Gregory Bateson, quoted in Mark Stahlman, "The Inner Senses and Human Engineering," *Dianoetikon* 1 (2020): 1–26.

104 **didn't come off as nefarious:** For the rich history of Bateson and Mead's efforts in this regard, see Fred Turner's terrific *The Democratic Surround*.

104 **Data scientists:** Jill Lepore, *If Then: How the Simulmatics Corporation Invented the Future* (New York: Liveright, 2020).

106 **"the purpose of Behavior Design":** Stanford University, "Welcome | Behavior Design Lab," https://captology.stanford.edu/, accessed June 18, 2018.

106 **Chatbots engage:** "Smartphone App to Change Your Personality," *Das Fachportal für Biotechnologie, Pharma und Life Sciences*, February 15, 2021, https://www .bionity.com/en/news/1169863/smartphone-app-to-change-your-personality .html.

107 **Amazon incentivizes productivity:** Nick Statt, "Amazon Expands Gamification Program That Encourages Warehouse Employees to Work Harder," *Verge*, March 15, 2021, https://www.theverge.com/2021/3/15/22331502/amazon -warehouse-gamification-program-expand-fc-games.

107 **promote environmentally friendly behavior:** Markus Brauer and Benjamin D. Douglas, "Gamification to Prevent Climate Change: A Review of Games and Apps for Sustainability," *Current Opinion in Psychology* 41 (December 1, 2021): 89–94, https://doi.org/10.31219/osf.io/3c9zj.

107 **Evgeny Morozov points out:** Evgeny Morozov, *To Save Everything, Click Here: The Folly of Technological Solutionism* (New York: PublicAffairs, 2013).

107 **Hook Model:** Nir Eyal, *Hooked: How to Build Habit-Forming Products* (New York: Portfolio, 2014).

108 **Red flags abound:** Andrea Valdez, "This Big Facebook Critic Fears Tech's Business Model," *Wired*, March 10, 2019, https://www.wired.com/story/this -big-facebook-critic-fears-techs-business-model/.

108 **"as big an existential threat":** *The Social Dilemma*, directed by Jeff Orlowski (Exposure Labs, The Space Program, Agent Pictures, 2020).

109 **"They're willing to see":** Douglas Rushkoff, interview with Naomi Klein, *Team Human* podcast, August 4, 2021, https://www.teamhuman.fm/episodes/ naomi-klein.

Chapter 9: Visions from Burning Man

112 *Shark Tank:* A reality show where entrepreneurs pitch billionaires including Mark Cuban and Barbara Corcoran for investment.

113 **"It's well documented":** Nellie Bowles, "'Burning Man for the 1%': The Des-

ert Party for the Tech Elite, with Eric Schmidt in a Top Hat," *Guardian*, May 2, 2016, https://www.theguardian.com/business/2016/may/02/further-future -festival-burning-man-tech-elite-eric-schmidt.

113   **"It's important":** Bowles, "'Burning Man for the 1%.'"

118   **funneling capital from Colombia:** Keith Larsen, "Investors Accuse Prodigy Network of Fraud at Troubled Park Ave Development," *Real Deal*, September 24, 2020, https://therealdeal.com/2020/09/24/investors-accuse-prodigy-network -of-fraud-at-troubled-park-ave-development/.

118   **charges of fraud:** Global Property and Asset Management Inc., "Panic at Prodigy," October 3, 2019, https://globalpropertyinc.com/2019/10/03/panic -at-prodigy/.

118   **"game of life":** "Akasha—The Game of Life," https://www.playakasha.com, accessed August 10, 2021.

119   **"exponential technologies . . . moonshots":** Singularity University, "An Exponential Primer," https://su.org/concepts/, accessed August 10, 2021.

119   **"entrepreneurial leaders . . . planetary scale":** Singularity University, "Singularity University," https://su.org/, accessed August 10, 2021.

120   **MacArthur Foundation:** MacArthur Foundation, "100 & Change," https:// www.macfound.org/programs/100change/.

122   **She hated Robert Moses's:** Jane Jacobs, *Systems of Survival: A Dialogue on the Moral Foundations of Commerce and Politics* (New York: Random House, 1992).

123   **new urbanism now amounts to:** Rushkoff, *Life Inc.: How Corporatism Conquered the World, and How We Can Take It Back* (New York: Random House, 2009), 74–83.

124   **Rutt has applied:** Jim Rutt, "A Journey to Game B," *Medium*, January 14, 2020, https://medium.com/@memetic007/a-journey-to-gameb-4fb13772bcf3.

125   **President Eisenhower:** Center for the Study of Digital Life, http://digitallife .center/, accessed August 10, 2021.

125   **"Yet in holding scientific":** Dwight D. Eisenhower, "Farewell Address to the Nation," available at http://mcadams.posc.mu.edu/ike.htm.

125   **white male–dominated tech industry:** S. Bodker and J. Greenbaum, "Design of Information Systems: Things versus People," in *Gendered Design: Information Technology and Office Systems*, ed. J. Owen Green and D. Pain (London: Taylor and Francis, 1993).

125   **"they would be based":** Sude V. Rosser, "Through the Lenses of Feminist Theory: Focus on Women and Information Technology," *Frontiers: A Journal of Women Studies* 26, no. 1 (2005).

125   **totalitarian surveillance state:** Alexandra Ma, "China's 'Social Credit' System Ranks Citizens and Punishes Them with Throttled Internet Speeds and Flight Bans If the Communist Party Deems Them Untrustworthy," *Business Insider*,

May 9, 2021, https://www.businessinsider.com/china-social-credit-system-punishments-and-rewards-explained-2018-4.

126   **put Black people in jail for longer:** See Cathy O'Neil, *Weapons of Math Destruction: How Big Data Increases Inequality and Threatens Democracy* (New York: Broadway Books, 2016).

127   **only a few hundred thousand had been shipped:** Namank Shah, "A Blurry Vision: Reconsidering the Failure of the One Laptop Per Child Initiative," *WR: Journal of the CAS Writing Program*, https://www.bu.edu/writingprogram/journal/past-issues/issue-3/shah/.

127   **"that are inappropriate":** Shah, "A Blurry Vision."

127   **could play only Western beats:** Victoria McArthur, "Communication Technologies and Cultural Identity: A Critical Discussion of ICTs for Development," paper presented at the IEEE Toronto International Conference: Science and Technology for Humanity, 2009, 910–14.

127   **$40 million on Brigade:** Micah Sifry, "Parker Bros," *Civicist*, February 12, 2019, https://civichall.org/civicist/parker-bros/.

127   **hub for planning civic technology projects:** Josh Constine, "Sean Parker's Govtech Brigade Breaks Up, Pinterest Acquires Engineers," *TechCrunch*, February 11, 2019, https://techcrunch.com/2019/02/10/brigade-pinterest/.

127   **"a big fat nothingburger":** Micah Sifry, email interview with the author, May 27, 2021.

128   **"submission of all forms of cultural life":** Neil Postman, *Technopoly: The Surrender of Culture to Technology* (New York: Vintage, 1993).

Chapter 10: The Great Reset

132   **The Covid pandemic:** "Covid Vaccines Create 9 New Billionaires with Combined Wealth Greater than Cost of Vaccinating World's Poorest Countries," Oxfam International, September 2, 2021, https://www.oxfam.org/en/press-releases/covid-vaccines-create-9-new-billionaires-combined-wealth-greater-cost-vaccinating.

134   **Solar panel disposal:** Maddie Stone, "Solar Panels Are Starting to Die, Leaving Behind Toxic Trash," *Wired*, August 22, 2020, https://www.wired.com/story/solar-panels-are-starting-to-die-leaving-behind-toxic-trash/.

136   **"If no one power":** Klaus Schwab and Thierry Malleret, *Covid-19: The Great Reset* (Cologny, Switzerland: Forum, 2020), 104.

137   **mosquito nets:** Stockholm University, "Mosquito nets: Are they catching more fishes than insects?," *ScienceDaily*, www.sciencedaily.com/releases/2019/11/191111100910.htm; Jeffrey Gettleman, "Meant to Keep Malaria Out, Mosquito Nets Are Used to Haul Fish In," *New York Times*, Janu-

ary 24, 2015, https://www.nytimes.com/2015/01/25/world/africa/mosquito
-nets-for-malaria-spawn-new-epidemic-overfishing.html.

138 **lifetime carbon footprint:** Katarina Zimmer and Carl-Johan Karlsson, "Green
Energy's Dirty Side Effects," *Foreign Policy*, June 18, 2020, https://foreignpolicy
.com/2020/06/18/green-energy-dirty-side-effects-renewable-transition
-climate-change-cobalt-mining-human-rights-inequality/.

139 **For renewables to provide:** "Earth and Humanity: Myth and Reality," YouTube video, May 16, 2021, 2:52:14, https://www.youtube.com/
watch?v=qYeZwUVx5MY.

139 **Replacing a majority:** "To replace all UK-based vehicles today with electric
vehicles, assuming they use the most resource-frugal next-generation NMC
811 batteries, would take 207,900 tonnes of cobalt, 264,600 tonnes of lithium
carbonate (LCE), at least 7,200 tonnes of neodymium and dysprosium, in addition to 2,362,500 tonnes of copper. This represents just under two times the
total annual world cobalt production, nearly the entire world production of
neodymium, three quarters of the world's lithium production and at least half
of the world's copper production during 2018." Energy and Our Future, "Earth
and Humanity: Myth and Reality," YouTube video, May 16, 2021, 2:52:14,
https://www.youtube.com/watch?v=qYeZwUVx5MY.

139 **Transitioning slowly:** Richard Heinberg, *Power: Limits and Prospects for Human
Survival* (Gabriola, BC, Canada: New Society, 2021).

139 **Degrowth is the only surefire way:** For more on degrowth, see the books and
resources listed on the Post Carbon Institute website, https://www.postcarbon
.org/.

140 **the worst accusations about these people:** For more, see Whitney Webb, *One
Nation Under Blackmail* (Chicago: Trine Day, 2022); Whitney Webb, "The Cover-
Up Continues: The Truth About Bill Gates, Microsoft, and Jeffrey Epstein,"
Unlimited Hangout, July 24, 2021, https://unlimitedhangout.com/2021/05/
investigative-reports/the-cover-up-continues-the-truth-about-bill-gates
-microsoft-and-jeffrey-epstein/.

140 **"at the forefront":** Steven Levy, "Bill Gates and President Bill Clinton on the
NSA, Safe Sex, and American Exceptionalism," *Wired*, November 12, 2013.

141 **Funders, scientists, and royals:** Whitney Webb, "Isabel Maxwell: Israel's
'Back Door' into Silicon Valley," *Unlimited Hangout*, July 24, 2021, https://
unlimitedhangout.com/2020/07/investigative-reports/isabel-maxwell
-israels-back-door-into-silicon-valley/; Naomi Klein, *The Shock Doctrine: The
Rise of Disaster Capitalism*, (Toronto: Alfred A. Knopf Canada, 2007); Anand
Giridharadas, *Winners Take All: The Elite Charade of Changing the World*, (New
York: Penguin, 2018); CNN staff, "Clinton's Foundation Got Millions from
Saudis, Gates," CNN politics, December 18, 2008 https://www.cnn.com/2008/

POLITICS/12/18/clinton.donations/; Leslie Sanchez, "Thorny Thicket of Bill and Hillary Clinton Conflicts?" CNN Politics, December 3, 2008. https://www.cnn.com/2008/POLITICS/12/03/sanchez.clinton/index.html.

141   **Do just a little reading:** Emily Flitter and James B. Stewart, "Bill Gates Met with Jeffrey Epstein Many Times, Despite His Past," *New York Times*, October 12, 2019, https://www.nytimes.com/2019/10/12/business/jeffrey-epstein-bill-gates.html; Tom Sykes, "Prince Andrew Was 'Given' 'Beautiful Young Neurosurgeon' by Epstein, Says Ex-Housekeeper," *The Daily Beast*, November 22, 2019, https://www.thedailybeast.com/prince-andrew-was-given-beautiful-young-neurosurgeon-by-jeffrey-epstein-says-ex-housekeeper; Kate Briquelet, "Melinda Gates Warned Bill About Jeffrey Epstein," *The Daily Beast*, May 7, 2021; Humanity+, "Humanity+ Clarification of Epstein Donation," accessed August 10, 2021, https://web .archive .org/web/20210808214020/https://humanityplus .org/humanity -clarification -of -epstein -donation/; "Sustainable Oceans Alliance: Impacting The SGDs," *Clinton Foundation*, December 22, 2016, https://www .clintonfoundation .org/clinton -global -initiative/commitments/sustainable -oceans -alliance -impacting -sgds; Jacob Bernstein, "Whatever Happened to Ghislaine Maxwell's Plan to Save the Oceans?," *The New York Times*, August 14, 2019, https://www.nytimes.com/2019/08/14/style/ghislaine-maxwell-terramar-boats-jeffrey-epstein.html.

141   **For this global oligarchy:** Amy Julia Harris, Frances Robles, Mike Baker, and William Rashbaum, "How a Ring of Women Allegedly Recruited Girls for Jeffrey Epstein," *New York Times*, August 29, 2019, https://www.nytimes.com/2019/08/29/nyregion/jeffrey-epstein-ghislaine-maxwell.html.

141   **Bill Gates has employed this logic:** "Bill Gates Buys Big on a Farmland Shopping Spree," *DW*, https://www.dw.com/en/bill-gates-buys-big-on-a-farmland-shopping-spree/a-57134690.

142   **"civilizational collapse":** Sissi Cao, "Bill Gates' Comments On COVID-19 Vaccine Patent Draw Outrage," *Observer*, April 27, 2021, https://observer.com/2021/04/bill-gates-oppose-lifting-covid-vaccine-patent-interview/.

142   **"despite his cuddly reputation":** Cory Doctorow, "Manufacturing MRNA Vaccines Is Surprisingly Straightforward," *Medium*, May 6, 2021, https://coronavirus.medium.com/manufacturing-mrna-vaccines-is-surprisingly-straightforward-despite-what-bill-gates-thinks-222cffb686ee.

142   **Gates argued:** Doctorow, "Manufacturing MRNA Vaccines Is Surprisingly Straightforward."

142   **new mRNA vaccines:** Kis, Zoltán, Cleo Kontoravdi, Antu K. Dey, Robin Shattock, and Nilay Shah, "Rapid Development and Deployment of High-Volume Vaccines for Pandemic Response," *Journal of Advanced Manufacturing and Processing* 2, no. 3 (2020), https://doi.org/10.1002/amp2.10060.

143   **Over Gates's objections:** Cao, "Bill Gates' Comments On COVID-19 Vaccine Patent Draw Outrage."

Chapter 11: The Mindset in the Mirror

144   **Klaus Schwab's vision:** Klaus Schwab and Thierry Malleret, *Covid-19: The Great Reset* (Cologny, Switzerland: Forum, 2020).

145   **"fixed traditional sentiments":** Robbie Shilliam, "How Black Deficit Entered the British Academy," https://robbieshilliam.wordpress.com/2017/06/20/how-black-deficit-entered-the-british-academy/, retrieved June 19, 2019.

146   **At the university level:** Adam Kirsch, "Technology Is Taking Over English Departments," *New Republic*, May 2, 2014, https://newrepublic.com/article/117428/limits-digital-humanities-adam-kirsch.

148   **Feeling blamed for society's ills:** Adi Robertson, "The FBI has released its Gamergate investigation records," *Verge*, January 27, 2017, https://www.theverge.com/2017/1/27/14412594/fbi-gamergate-harassment-threat-investigation-records-release.

148   **Gamergate:** Michael James Heron, Pauline Belford, and Ayse Goker, "Sexism in the Circuitry," *ACM SIGCAS Computers and Society* 44, no. 4: 18–29, https://doi.org/10.1145/2695577.2695582.

149   **"He wanted to destroy":** Ronald Rodash, "Steve Bannon, Trump's Top Guy, Told Me He Was 'a Leninist,'" *Daily Beast*, April 13, 2017, https://www.thedailybeast.com/steve-bannon-trumps-top-guy-told-me-he-was-a-leninist.

149   **Bannon may believe:** Jeremy W. Peters, "Bannon's Worldview: Dissecting the Message of 'The Fourth Turning,'" *New York Times*, April 8, 2017, https://www.nytimes.com/2017/04/08/us/politics/bannon-fourth-turning.html.

149   **1960s science fiction novel:** Roger Zelazny, *Lord of Light* (New York: Harper Voyager, 2010).

149   **"In Silicon Valley":** Andy Beckett, "Accelerationism: How a Fringe Philosophy Predicted the Future We Live In," *Guardian*, May 11, 2017, https://www.theguardian.com/world/2017/may/11/accelerationism-how-a-fringe-philosophy-predicted-the-future-we-live-in.

150   **"It's a fine line":** Max Chafkin, *QAnon Anonymous* podcast, December 10, 2021.

150   **"cognitive elite":** Mark O'Connell, "Why Silicon Valley Billionaires Are Prepping for the Apocalypse in New Zealand," *Guardian*, February 15, 2018, https://www.theguardian.com/news/2018/feb/15/why-silicon-valley-billionaires-are-prepping-for-the-apocalypse-in-new-zealand.

150   **Thiel also funded:** Max Chafkin, *The Contrarian: Peter Thiel and Silicon Valley's Pursuit of Power* (New York: Penguin, 2021).

Chapter 12: Cybernetic Karma

159   **cybernetics:** Norbert Wiener, *Cybernetics: Or Control and Communication in the Animal and the Machine* (New York: Wiley, 1948).

160   **"a kind of vaccination":** Nora Bateson, *Small Arcs of Larger Circles: Framing Through Other Patterns* (Charmouth, UK: Triarchy Press, 2016), 198–99.

161   **butterfly flapping:** Edward Lorenz, speech to the American Association for the Advancement of Science, Washington, DC, December 29, 1972, transcribed in Edward Lorenz, *The Essence of Chaos* (Seattle: University of Washington Press, 1993).

162   **"anyone may publish":** Ken Jordan and Randall J. Packer, *Multimedia: From Wagner to Virtual Reality* (New York: W. W. Norton, 2001).

162   **"We become what we behold":** J. M. Culkin, "A Schoolman's Guide to Marshall McLuhan," *Saturday Review*, March 18, 1967, 51–53, 71–72.

164   **I wrote my dissertation:** Douglas Rushkoff, "Monopoly Moneys," PhD diss., Utrecht University, 2012.

164   **the more frequently retail traders transacted:** Dalbar, Inc., "Quantitative Analysis of Investor Behavior 2011" (Boston: Dalbar, Inc., 2011).

165   **The stock shot upwards:** Eric Lam and Lu Wang, "Steely Meme-Stock Short Sellers Stare Down $4.5 Billion Loss," *Bloomberg*, June 3, 2021, https://www.bloomberg.com/news/articles/2021-06-03/defiant-meme-stock-short-sellers-stare-down-4-5-billion-loss.

166   **A platform like TikTok:** Shelly Banjo and Shawn Wen, "A Push-Up Contest on TikTok Exposed a Great Cyber-Espionage Threat," *Bloomberg*, May 13, 2021, https://www.bloomberg.com/news/articles/2021-05-13/how-tiktok-works-and-does-it-share-data-with-china.

167   **"They all know the algorithms":** Taylor Lorenz, Kellen Browning, and Sheera Frenkel, "TikTok Teens and K-Pop Stans Say They Sank Trump Rally," *New York Times*, June 21, 2020, https://www.nytimes.com/2020/06/21/style/tiktok-trump-rally-tulsa.html.

167   **formed a union:** Zoe Schiffer, "Exclusive: Google Workers across the Globe Announce International Union Alliance to Hold Alphabet Accountable," *Verge*, January 25, 2021, https://www.theverge.com/2021/1/25/22243138/google-union-alphabet-workers-europe-announce-global-alliance.

167   **"sometimes the boss is the best organizer":** Kate Conger, "Hundreds of Google Employees Unionize, Culminating Years of Activism," *New York Times*, January 4, 2021, https://www.nytimes.com/2021/01/04/technology/google-employees-union.html.

169   **an open letter about the frightening potential:** Wikimedia, "Open Letter on Artificial Intelligence," https://en.wikipedia.org/wiki/Open_Letter_on_Artificial_Intelligence, accessed August 10, 2021.

170 **"Things are getting . . . currently doing":** Cat Clifford, "Billionaire Tech Titan Mark Cuban on AI: 'It Scares the S— Out of Me,'" *CNBC*, July 25, 2017, https://www.cnbc.com/2017/07/25/mark-cuban-on-ai-it-scares-me.html.

170 **"Is the country going to turn":** Evan Osnos, "Doomsday Prep for the Super Rich," *New Yorker*, January 22, 2017, https://www.newyorker.com/magazine/2017/01/30/doomsday-prep-for-the-super-rich.

170 **Employees protested:** Peter Kafka, "Google Wants out of the Creepy Military Robot Business," *Vox*, March 17, 2016, https://www.vox.com/2016/3/17/11587060/google-wants-out-of-the-creepy-military-robot-business.

170 **four thousand Googlers:** Kate Conger, "Google Employees Resign in Protest Against Pentagon Contract," *Gizmodo*, May 14, 2018, https://gizmodo.com/google-employees-resign-in-protest-against-pentagon-con-1825729300.

171 **"the one who becomes the leader":** Associated Press, "Putin: Leader in Artificial Intelligence Will Rule World," *CNBC*, September 4, 2017, https://www.cnbc.com/2017/09/04/putin-leader-in-artificial-intelligence-will-rule-world.html.

171 **"I think the danger of AI":** Elon Musk Answers Your Questions! | SXSW 2018," YouTube video, posted by South by Southwest, March 11, 2018, 1:11:37, https://www.youtube.com/watch?v=kzlUyrccbos.

171 **"Whereas the short-term impact . . . with AI":** Stephen Hawking, Stuart Russell, Max Tegmark, Frank Wilczek, "Stephen Hawking: 'Transcendence Looks at the Implications of Artificial Intelligence—But Are We Taking AI Seriously Enough,'" *Independent*, October 23, 2017, https://www.independent.co.uk/news/science/stephen-hawking-transcendence-looks-implications-artificial-intelligence-are-we-taking-ai-seriously-enough-9313474.html.

173 **"It's not that I hate AI":** Similarly, Roko's basilisk, a thought experiment initiated on a discussion board for rational thought about technology called Less-Wrong, theorized that a future AI would have an incentive to torture anyone who could imagine the agent but did not try to help it come into existence. The argument was called a basilisk because anyone who heard it would forever more be at risk from the AI once it learned to time-travel. The conversation itself was banned for fear of spreading information hazards. "Roko's Basilisk," LessWrong, https://www.lesswrong.com/tag/rokos-basilisk, accessed December 9, 2021.

173 **"summoning the demon":** "Elon Musk at the MIT AeroAstro Centennial Symposium," YouTube Video, July 5, 2015, 1:23:27, https://www.youtube.com/watch?v=4DUbiCQpw_4.

173 **"so that we'll have a bolt-hole":** Maureen Dowd, "Elon Musk's Billion-Dollar Crusade to Stop the AI Apocalypse," *Vanity Fair*, March 26, 2017, https://www.vanityfair.com/news/2017/03/elon-musk-billion-dollar-crusade-to-stop-ai-space-x.

173   **Musk has been developing a neural net apparatus:** Dowd, "Elon Musk's Billion-Dollar Crusade."

Chapter 13: Pattern Recognition

176   **"To the Planetarium":** Walter Benjamin, "To the Planetarium," in *One-Way Street: And Other Writings*, translated by Edmund Jephcott (Brooklyn, NY: Verso, 2021).

179   **"You can build":** Tyson Yunkaporta, *Sand Talk: How Indigenous Thinking Can Save The World* (New York: HarperOne, 2020), 78–79.

179   **eating local foods is better for our health:** Vicki Robin, *Blessing the Hands That Feed Us: What Eating Closer to Home Can Teach Us about Food, Community, and Our Place on Earth* (Farming Hills, MI: Thorndike Press, 2014).

181   **We consume over three billion gallons:** Koustav Samanta and Roslan Khasawneh, and Florence Tan, "APPEC-Global oil demand seen reaching pre-pandemic levels by early 2022," *Reuters*, September 27, 2021, https://www.reuters.com/business/energy/appec-global-oil-demand-seen-reaching-pre-pandemic-levels-by-early-2022-2021-09-27/.

183   **They pooled money:** Jessica Gordon Nembhard, *Collective Courage: A History of African American Cooperative Economic Thought and Practice* (University Park, PA: Pennsylvania State University Press, 2014).

186   **"Young people feel":** Elise Chen, "These Chinese Millennials Are 'Chilling,' and Beijing Isn't Happy," *New York Times*, July 3, 2021, https://www.nytimes.com/2021/07/03/world/asia/china-slackers-tangping.html.

186   **"Amidst global shutdown":** Gaya Herrington, "Beyond Growth," *WEFLIVE*, January 23, 2020, https://www.weflive.com/story/e968fb0963974e1e8f6c636e5654cbc2.

186   **"resource scarcity has not":** Edward Helmore, "Yep, It's Bleak, Says Expert Who Tested 1970s End-of-the-World Prediction," *Guardian*, July 25, 2021, https://www.theguardian.com/environment/2021/jul/25/gaya-herrington-mit-study-the-limits-to-growth.

188   **"only now . . . all of us":** *American Utopia*, directed by Spike Lee (HBO, 2021), https://www.hbo.com/specials/american-utopia.

188   **There's no "solution":** See Sarah Pessin's work, including "From Mystery to Laughter to Trembling Generosity: Agono-Pluralistic Ethics in Connolly v. Levinas," *International Journal of Philosophical Studies* 24, no. 5 (2016): 615–38.